Synthesis Lectures on Engineering, Science, and Technology

The focus of this series is general topics, and applications about, and for, engineers and scientists on a wide array of applications, methods and advances. Most titles cover subjects such as professional development, education, and study skills, as well as basic introductory undergraduate material and other topics appropriate for a broader and less technical audience.

Rafael Martínez-Guerra •
Juan Javier Montesinos-García •
Juan Pablo Flores-Flores

Encryption and Decryption Algorithms for Plain Text and Images using Fractional Calculus

 Springer

Rafael Martínez-Guerra
Automatic Control
Center for Research and Advanced Studies of
the National Polytechnic Institute
(CINVESTAV-IPN)
Mexico City, Mexico

Juan Javier Montesinos-García
Institute of Electronics and Mechatronics
Technological University of the Mixteca
Huajuapan de León, Mexico

Juan Pablo Flores-Flores
Automatic Control
Center for Research and Advanced Studies of
the National Polytechnic Institute
(CINVESTAV-IPN)
Mexico City, Mexico

ISSN 2690-0300   ISSN 2690-0327   (electronic)
Synthesis Lectures on Engineering, Science, and Technology
ISBN 978-3-031-20700-6   ISBN 978-3-031-20698-6   (eBook)
https://doi.org/10.1007/978-3-031-20698-6

This Springer imprint is published by the registered company Springer Nature Switzerland AG
The registered company address is: Gewerbestrasse 11, 6330 Cham, Switzerland

In memory of my father,
Carlos Martínez Rosales.
To my wife and sons,
Marilen, Rafael, and Juan Carlos.
To my mother and brothers,
Virginia, Victor, Arturo, Carlos,
Javier, and Marisela.

Rafael Martínez-Guerra

To my mother Irma and my father Javier,
whose love and support allowed me
to reach all my goals.
To my aunt Rosario and Uncle Clemente,
their teachings, love and support were
instrumental in my formation up to today.
To my cousin Oscar, for all these years
of support and assertive counsel.
In loving memory of my grandmother Sofia,
aunt Margarita and cousin Daniel.

Juan Javier Montesinos-García

To my family,
Eladio, Candelaria, Edy, Abril,
Mario, Amaia, Akane and Najmeh.

Juan Pablo Flores-Flores

Preface

Many people know that it is possible to intercept and modify data if an application does not protect it when travels on an untrusted network, and the application then becomes a disaster when it comes to security. In this book, we offer an alternative to encrypt and decrypt messages using objects called integer and fractional-order estimators or observers, by means of security codes. We establish a class of observers capable of carrying out this work, by means of security codes where finally, since an observer is nothing more than a mathematical model represented through nonlinear differential equations that can be of integer or fractional type that serve as means to send messages either of the plain-text type or of the image type whose key or security code to encrypt or decrypt is nothing more than a set of initial conditions where it makes sense to speak of this means of transporting the message either plain-text or image for specific attacks for chaotic cryptosystems of the stream cipher type. In this book, we mention the type of observers to treat either the integer or fractional order type and their main characteristics. We discuss an important property of some systems such as Liouville systems that is very important for the encryption and decryption of messages in integer and fractional order nonlinear systems by using the synchronization property of chaotic systems where we address some logistic maps such as Mandelbrot sets including Julia and fractal sets taking advantage of their characteristics to encrypt or recover messages. We discuss some issues about stream and block ciphers and some state observers. Various types of observers are proposed for nonlinear systems of integer and fractional order from the simplest (Luenberger Observer) to the most sophisticated such as the Supertwisting Observer for message receivers as well as their vulnerability to attacks. Observers of the exponential polynomial type are proposed together with the property of the Liouville type. We also propose the usefulness of robust fractional systems of sliding modes with Liouville characteristics as means of transmission and reception of plain-text and image messages. Of all the alternatives for encryption and decryption of messages shown here, a vulnerability analysis to cryptographic attacks (cryptoanalysis) is made, this is a security analysis, an important topic on the subject of secure communications. The book is self-contained, that is to say, the necessary tools to address the issues such as fractional calculus are given in the same book and several examples are presented. Moreover, this book includes exercises that are left to the reader. The book is directed to an audience such as professionals in the areas of mathematics, physics and engineering and researchers in general and related areas with a minimum of knowledge in higher mathematics. However, it also contains advanced research topics for people interested in encryption and decryption, observers, synchronization and secure communications areas. The book is organized as follows. In Chap. 1, a brief overview of the main topics covered is presented giving an introduction to the state of the art on encryption and decryption algorithms, synchronization of chaotic systems, security keys or codes, security analysis such as cryptographic attacks, linear and differential cryptanalysis, in addition to specific attacks for chaotic cryptosystems of type stream cipher. In Chap. 2, some definitions are given about the Lyapunov exponents, stability, and state observers; also fractals and synchronization are briefly introduced. Chapter 3 shows the stream and block ciphers and observers, binary representations as well as some conversions from binary to decimal and vice versa, representations of plain text and images in integer bits and ciphers with generalized synchronization. Chapter 4 deals with the study of Liouville systems and cryptography, and a supertwisting observer is addressed as a receiver as well as its vulnerability to cryptanalysis. Chapter 5 presents some basic concepts of state observers, the exponential polynomial observer is used as a receptor, and the receptors are based on properties related to Liouville systems. Chapter 6 shows some basic elements of fractional calculus and some observers. Chapter 7 deals with the implementation of systems with the property of Liouville and fractional order systems used for the encryption and decryption of plain-text and image messages. In Chap. 8, we present robust fractional order state observers as means of encryption and decryption, presenting a security analysis and situations that lead to decryption failures. Finally, in Chap. 9, a new topic is described in secure communications, and we present encryption and decryption algorithms by using state observers that are represented by means of fractional-order chaotic systems with the Atangana-Baleunu fractional derivative. Additionally, the reader will find throughout this material some exercises to strengthen the knowledge acquired.

Mexico City, Mexico   Rafael Martínez-Guerra
                                                                              Juan Javier Montesinos-García
                                                                                 Juan Pablo Flores-Flores

Contents

Notations and Abbreviations

$\mathbb{N}$	The set of natural numbers
$\mathbb{Z}$	The set of integers numbers
$\mathbb{Q}$	The set of rational numbers
$\mathbb{R}$	The set of real numbers
$\mathbb{C}$	The set of complex numbers
$A, B, \ldots$	Capital letters represent arbitrary sets
$x, y, \ldots$	Lowercase letters represent elements of a set
$A \subset B$	A is subset of B
$x \in A$	x is element of A
$A \cup B$	The union of two sets
$A \cap B$	The intersection of two sets
A^c	Complement set of A
$A \setminus B$	Difference of sets A and B
$\oplus$	Bitwise XOR logical operation
$\emptyset$	Empty set
$\Longleftrightarrow$	Necessary and sufficient condition
$\forall$	For all
$\sim$	Equivalence relation
$a \equiv b \mod n$	a is congruent with b module n
$*$	Binary operation for groups
$\det(A)$	The determinant of a square matrix $A \in \mathbb{R}^{n \times n}$
$\lvert A \rvert$	The determinant of a square matrix $A \in \mathbb{R}^{n \times n}$
$\mathrm{Tr}(A)$	The trace of a matrix A
A^{T}	The transpose of a matrix A
$\{\}$	Set
$(a_{ij})_{i,j}$	$m \times n$ matrix with entries a_{ij}, $1 \le i \le m, 1 \le j \le n$
$\mathrm{rank}(A)$	Rank of a matrix A
A^{-1}	Inverse of A
$\dot{y} = \frac{dy}{dt}$	First derivative of y with respect to t

$\square$	Designation of the end of a proof
$< (>)$	Less (greater) than
$\leq (\geq)$	Less (greater) than or equal to

List of Figures

Introduction

1

Abstract

In this chapter we give an overview of cryptography and cryptanalysis where basic concepts and definitions are given, also the relation of cryptography to chaotic systems synchronization is addressed.

1.1 Chaotic System Synchronization and Encryption Algorithms

Chaotic systems synchronization was introduced in [1]. There is proposed a methodology such that a chaotic system called slave, follows the state trajectories of a second chaotic system called master. This, by means of a coupling signal. Subsequently, numerous proposals have emerged to achieve the same goal, such as complete synchronization, generalized synchronization, impulsive synchronization, phase synchronization, delay synchronization, etc. As a result, multiple applications of chaotic systems synchronization have been found. One of the most important is secure communications [2–9], which is the main topic of this book.

Most encryption algorithms based on chaotic systems can be classified into one of the following kinds of encryption:

- **Chaotic masking**: It consists of adding the signal message on the output of a chaotic system.
- **Chaos shift keying (CSK)**: It consists of transmitting a message as variations of a given parameter of the chaotic system. It usually requires converting the message to its binary equivalent, therefore, in this scheme 1 corresponds to a specific parameter

value and 0 to another. As a result, changes occur in the behavior of the chaotic system attractor.

- **Chaotic modulation**: Here, the message changes the value of some parameter of the chaotic system.

These are the best known methods that consist in encrypting messages with chaotic systems and then recovering them by means of a synchronization. A more extensive explanation of these encryption algorithms, based on chaotic systems is given below.

1.1.1   Encryption Through Chaotic Systems

There exist two basic types of chaotic cryptosystems: analog and digital. The former are mainly based on the synchronization of chaotic systems, the second ones, can be independent of synchronization and are completely digital. The implementation of an analog encryption algorithm requires the circuits responsible of generating chaos to be described in sufficient detail, such as the explicit form of the differential equation of the system and parameters that generate chaotic behavior. Meanwhile, for digital systems, precision, arithmetic (floating or fixed point), hardware configuration, among others, must be given.

In general, encryption algorithms are typically divided into two kinds: symmetric key and asymmetric key. The former uses the same key to encrypt and decrypt information, as a consequence these are extremely fast and useful to handle large volumes of data at high speed. This kind of algorithms are divided into two classes:

- **Stream cipher**: These generate a pseudo-random stream of symbols (keystream) by means of a public deterministic algorithm, which is governed by a secret key. Thus, the message is combined with the keystream, usually with a two-module sum or with a bitwise XOR. Among the most common stream ciphers we have: A5 / 1, A5 / 2, E0. RC4, SEAL, etc.
- **Block encryption**: These encrypt the original message by clustering its elements in blocks of two or more, so that each encrypted block is of the same size. These algorithms usually consist of an initial transformation, a cryptographic function iterated certain number of times and a final transformation. Then, the key is expanded using some algorithm so that enough key elements are obtained for each round of encryption. The most popular algorithms of this kind are: AES, DES, RC5, etc.

Symmetric key algorithms generally have keys between 128 and 256 bits. On the other hand, asymmetric key algorithms use two keys for the encryption and decryption process. Usually one of the keys is public and the other private. When encrypting, both keys are used and to decrypt only the private key is necessary. These algorithms are generally slow as require complex operations with large integers, so they are used to encrypt small data packets such as digital signatures, secret key agreements, etc. The most common public key algorithm is RSA and its keys usually require between 1024 and 4096 bits [10].

The key is fundamental in any encryption algorithm. However, very few works in the literature, proportionate detailed information about the key. Therefore, in the following a simple introduction is given to understand the importance of these keys and how they should be obtained for chaotic cryptosystems.

1.2   Key or Security Code

A common element in all encryption algorithms is the key. The security code of an encryption algorithm must depend exclusively on the key [11].

No matter how strong and how well designed an algorithm is, if the key is not adequate, then the encryption will be easily violated [10]. As has been said, there is little information about how to choose or design the keys. Moreover, fundamental specifications such as the space of the key or the variables to be used as key are not presented. Therefore, here this will be one of the main aspects to be covered when proposing encryption algorithms.

The key space is defined as the set of values that can be used as keys. The size of the key space is given by the number of possible keys for the system. The keys from classical encryption algorithms are usually strings of random bits that are generated by some automatic process. In the chaotic encryption schemes, the properties of the key space elements are not the same, since not all keys are equally strong. For this, a bifurcation diagram can be useful to find intervals in which a key produces periodic orbits and thus avoid the use of weak keys (degenerate keys).

When many parameters are used simultaneously as a key, finding the most convenient intervals (without degenerate keys) might be difficult due to the dependence between the parameters. In such case, the positive Lyapunov exponents can be used to describe the key space. Thus, it must be obtained the largest Lyapunov exponent for the desired parameter combinations, then, if the obtained exponent is positive, the parameter combination can be used as a key.

The key space must be large enough to avoid brute force attacks. However, if the region that produces chaotic behavior is not large enough, it must be increased as much as possible to avoid equivalent keys. That is, to avoid a group of keys that can decrypt the same encrypted message due to its closeness. Thus, the region that produces chaotic behavior must be discretized so that the space between adjacent keys does not produce equivalent keys.

The keys should favor the presence of the so called avalanche property, that is, when a change occurs in the key the encrypted message will change radically and ideally it should change at least half of the values of the encrypted message. In some cases, the chaotic system parameters are set and only one of them is used as a key, which can be counterproductive since it is possible to use a bit-error-rate (BER) attack in which some system parameters are set and from these, an approximation of the one used as the key can be obtained. Therefore, the partial knowledge of the key should never reveal information about the message or the unknown part of the key.

The security of an encrypted message is usually given by the priority when designing an encryption algorithm. When a new encryption algorithm is presented, it is common to provide a security analysis of it. Therefore, in the following will be presented an explanation on what an appropriate security analysis should contain for each type of cryptosystem.

1.3   Security Analysis

Security is the main interest of an encryption algorithm, so it must be evaluated at least by a basic security analysis. This is, it must at least withstand the most known and popular attacks to identify and correct defects before the system is published.

The algorithm will be resistant to the most common attacks if have two basic characteristics: confusion and diffusion. The first makes the relationship between the key and the encrypted message as complex as possible, making difficult to find redundancies or statistical patterns in the encrypted message. The second property consists of rearranging or scattering the bits in a message so that the influence of the message and the key are dispersed as best as possible within the encrypted message. To fulfill these requirements, the algorithm must satisfy the following [12]

1. Sensitivity with respect to the key, that is, changing a single character of the key produces completely different encrypted messages when the algorithm is applied to the same message.
2. Sensitivity with respect to the message, that is, altering one bit of the message should create totally different encrypted messages.
3. Absence of patterns in the encrypted text.

The first two characteristics generate confusion, while the last one is responsible for providing diffusion.

1.3.1   Cryptographic Attacks (Cryptanalysis)

During the security analysis, must be carried out attacks that assume that the cryptanalyst knows the exact design of the algorithm and how it works. This is, everything about the algorithm, except the key, is known. This must be done since the algorithm is sold to several users and therefore, it is reasonable to assume that it will easily be stolen, compromising all the details of its operation. Thus, the security of the algorithm must depend only on its key and not in the secrecy of its operation.

A cryptographic system can be described by the following elements:

1. P is the set of possible messages.
2. C is the set of possible encrypted messages.
3. K is the space of the key.
4. e_k is the encryption algorithm for each element $k \in K$.
5. d_k is the corresponding decryption algorithm for the element $k \in K$ mentioned in the previous point.

The operation of an algorithm can be summarized as follows: given a message $x \in P$, this can be encrypted with a key $k \in K$ by using the encryption rule $e(x, k) = y$, $y \in C$. Meanwhile, the encrypted message y is decrypted by using the corresponding decryption rule $d(y, k) = x$, such that $d(e(x, k), k) = x$.

There are several kinds of attacks to carry out the cryptanalysis of an algorithm. The most popular ones are listed below, starting with the most complicated:

1. **Ciphertext only (encrypted message)**: The attacker knows one or more encrypted messages $y_1, y_2, \ldots, y_n \in C$.
2. **Known plaintext (known message)**: The attacker knows one or more messages $x_1, x_2, \ldots, x_n \in P$ and its corresponding encrypted message $y_1, y_2, \ldots, y_n \in C$.
3. **Chosen plaintext (chosen message)**: The attacker has temporary access to the encryption device and can choose some messages $x_1, x_2, \ldots, x_n \in P$ as well as obtain the encrypted messages $y_1, y_2, \ldots, y_n \in C$ that are generated.
4. **Chosen ciphertext (chosen encrypted message)**: The attacker has temporary access to the encryption device and can choose some encrypted messages $y_1, y_2, \ldots, y_n \in C$ as well as obtain the messages $x_1, x_2, \ldots, x_n \in P$ that are generated.

The objective of each of these attacks is to obtain the key k or some equivalent key that was used to encrypt the messages. In particular, the attacks of known and chosen message are very effective in the cryptanalysis of algorithms based on chaotic systems [13–16]. There exist other attacks that are less common. However, these have characteristics of the already mentioned above. In the case of block encryption algorithms, the analysis on the susceptibility to differential and linear cryptanalysis should be included.

1.3.2 Differential Cryptanalysis

Differential cryptanalysis was introduced by Guojie et al. [17] and is a variant of the chosen message attacks, which tries to find the key of an iterative encryption algorithm. It consists of analyzing the differences caused in encrypted messages when performing determined changes in the messages that generated them. These differences are used to determine the most probable key among all the possible keys. At the same time, the number of tests that would be done when implementing a brute force attack is reduced. Usually, the difference is chosen as the result of a bitwise XOR operation between the two unencrypted messages.

1.3.3   Linear Cryptanalysis

Linear cryptanalysis was introduced in [18] and is essentially a known message attack whose purpose is to generate a linear expression that approximates a certain block cipher. A linear expression for a given iteration will be an equation that is based on the module-two sum between the inputs and outputs of such iteration.

1.4   Specific Attacks for Stream Cipher-Type Chaotic Cryptosystems

There are several cryptanalysis forms for stream cipher encryption algorithms based on chaotic systems. These can be classified as follows:

1. Extraction of the signal from the message $s(t)$ of the transmitted signal $y(t)$.
2. Extraction of the signal that carries the data $c(t)$ and then remove it and retrieve the message $s(t)$.
3. Estimation of the transmitter's secret parameters to completely break the algorithm.
4. Brute force attacks.

Each of these analyses is explained in greater detail below.

1.4.1   Message Extraction

When using chaotic masking techniques, extracting the signal is possible if the message $s(t)$ is a periodic signal during a sufficient amount of time. Methods such as auto-correlation and cross-correlation analysis, spectral power analysis, filtering techniques, and generalized synchronization are usually used.

Power spectral analysis and filtering take advantage of the chaotic signals limitations, which are used to mask the message. The power spectrum of the message must be completely covered with the power spectrum of the chaotic signal that was used to mask it. However, several encryption algorithms fail at this, since the commonly used chaotic oscillators, such as Rössler, Lorenz, Chua, Duffing, etc., have a much lower density power than common messages. Therefore, these cannot support this type of filter-based attacks.

The generalized synchronization attack was introduced in [19]. This assumes that the attractor used is known, but the oscillator parameters are ignored. Its purpose is to reconstruct the signals used to hide the message and then access the signal that contains the message.

1.4.2 Parametric Estimation

Several chaos-based encryption schemes are not sensitive enough to variations in transmitter and receiver parameters, allowing similar parameters to be used for message retrieval.

Different methods can be used for this, for example, it is possible to solve the differential equations based on the signals that they emit. Also, the parameters can be estimated from a generalized synchronization scheme. In addition, some adaptive control techniques can be useful to find equivalent keys.

1.4.3 Brute Force Attacks

A brute force attack consists on testing all the possible keys. The effectiveness of this attack will depend on the size of the key space and the attacker's processing capacity. It is commonly considered that any space with less than 2^{100} elements it is not safe, although this number increases when the processing power is improved [10].

References

1. Pecora, L. M., & Carroll, T. L. (1990). Synchronization in chaotic systems. *Physical Review Letters A*, *64*, 821–824.
2. Kocarev, L., Halle, K. S., Eckert, K., Chua, L. O., & Parlitz, U. (1992). Experimental demonstration of secure communications via chaotic synchronization. *International Journal of Bifurcation and Chaos*, *2*(03), 709–713.
3. Liao, T. L., & Huang, N. S. (1999). An observer-based approach for chaotic synchronization with applications to secure communications. *IEEE Transactions on Circuits and Systems I: Fundamental Theory and Applications*, *46*(9), 1144–1150.
4. Cuomo, K. M., Oppenheim, A. V., & Strogatz, S. H. (1993). Synchronization of Lorenz-based chaotic circuits with applications to communications. *IEEE Transactions on Circuits and Systems II: Analog and Digital Signal Processing*, *40*(10), 626–633.
5. Smaoui, N., Karouma, A., & Zribi, M. (2011). Secure communications based on the synchronization of the hyperchaotic Chen and the unified chaotic systems. *Communications in Nonlinear Science and Numerical Simulation*, *16*(8), 3279–3293.
6. Wang, S., Kuang, J., Li, J., Luo, Y., Lu, H., & Hu, G. (2002). Chaos-based secure communications in a large community. *Physical Review E*, *66*(6), 065202.
7. Li, Z., & Xu, D. (2004). A secure communication scheme using projective chaos synchronization. *Chaos, Solitons & Fractals*, *22*(2), 477–481.
8. Nana, B., Woafo, P., & Domngang, S. (2009). Chaotic synchronization with experimental application to secure communications. *Communications in Nonlinear Science and Numerical Simulation*, *14*(5), 2266–2276.

9. Li, C., Liao, X., & Wong, K. W. (2004). Chaotic lag synchronization of coupled time-delayed systems and its applications in secure communication. *Physica D: Nonlinear Phenomena, 194*(3–4), 187–202.
10. Schneier, B. (2007). *Applied cryptography: Protocols, algorithms, and source code in C.* John Wiley & Sons.
11. Menezes, A. J., Van Oorschot, P. C., & Vanstone, S. A. (1996). *Handbook of applied cryptography.* CRC Press.
12. Alvarez, G., & Li, S. (2006). Some basic cryptographic requirements for chaos-based cryptosystems. *International Journal of Bifurcation and Chaos, 16*(08), 2129–2151.
13. Biham, E., & Shamir, A. (2012). *Differential cryptanalysis of the data encryption standard.* Springer Science & Business Media.
14. Alvarez, G., Montoya, F., Romera, M., & Pastor, G. (2000). Cryptanalysis of a chaotic encryption system. *Physics Letters A, 276*(1–4), 191–196.
15. Stojanovski, T., Kocarev, L., & Parlitz, U. (1996). A simple method to reveal the parameters of the Lorenz system. *International Journal of Bifurcation and Chaos, 6*(12b), 2645–2652.
16. Li, C., Li, S., Zhang, D., & Chen, G. (2005, May). Chosen-plaintext cryptanalysis of a clipped-neural-network-based chaotic cipher. In *International Symposium on Neural Networks* (pp. 630–636). Berlin, Heidelberg: Springer.
17. Guojie, H., Zhengjin, F., & Ruiling, M. (2003). Chosen ciphertext attack on chaos communication based on chaotic synchronization. *IEEE Transactions on Circuits and Systems I: Fundamental Theory and Applications, 50*(2), 275–279.
18. Matsui, M. (1993, May). Linear cryptanalysis method for DES cipher. In *Workshop on the Theory and Application of of Cryptographic Techniques* (pp. 386–397). Berlin, Heidelberg: Springer.
19. Yang, T., Yang, L. B., & Yang, C. M. (1998). Breaking chaotic switching using generalized synchronization: Examples. *IEEE Transactions on Circuits and Systems I: Fundamental Theory and Applications, 45*(10), 1062–1067.

Synchronization of Chaotic Systems

2

Abstract

This chapter introduces important concepts about chaotic systems and how to determine if a system is chaotic by Lyapunov exponents, it also gives an introduction to state observers to synchronize two chaotic system, the observers stability and convergence is analyzed using the previously given theory on stability. The chapter concludes with a section introducing fractals and synchronization of fractals using dynamic control

2.1 Chaotic Systems

Most dynamic systems exhibit trajectories that after a transitory time converge to a certain equilibrium point.

Some dynamic systems have unstable dynamics during this transitory time and in some cases even show exponential separation between its trajectories, this is that the distance between the states increases at an exponential rate after every measure, hence the distance between them increases, but if it nears a sink point it is attracted to it and after some time it will converge to the sink, this is a common occurrence, where unstable dynamics are just transitory.

A chaotic trajectory is one that remains bounded and oscillates, this is that it does not diverge but neither converges to a sink point or periodic orbit, showing a behavior that seems random, at any moment there are points at the orbit that separate from its initial condition as time passes, these points are called Lyapunov numbers and Lyapunov exponents [1, 2].

A chaotic system is one where its states have chaotic trajectories, two initial points (initial conditions) arbitrarily close from each other at the beginning separate as time passes, never to reach a common sink or orbit.

2.1.1 Lyapunov Exponents

Chaotic trajectories separate from each other as time passes, not every dynamic systems has this property and in the ones that have it, not every trajectory can do this, the chaotic motion happens depending on the initial condition and parameters of the system, so it is necessary to measure if the distance between trajectories grows to determine the set of values that promote chaotic behavior.

The Lyapunov number is used to better measure the rate of separation of very close points, the Lyapunov exponent is the natural logarithm of the Lyapunov number, this means that the trajectories that surround the fixed point will separate from the point in average a certain number of times each iteration [3].

Consider the state $F_T(x_0)$ is the point where the trajectory with initial condition x_0 converges after t time units, let $\dot{x} = f(x_0)$ be a system of differential equations with $x = (x_1, \ldots, x_n)$ states, the Lyapunov exponent is defined as follows:

Definition 2.1 The Lyapunov numbers and exponents of a trajectory are defined as the Lyapunov numbers and exponents of the associated time map.

It is necessary to obtain the derivative of the map $DF_t(x_0)$ with respect of the initial condition v, if the time is fixed into one desired point, the derivative $DF_t(v)$ is a linear map, then

$$\frac{d}{dt}F_t(x_0) = f[F_t(x_0)]$$

By differentiating with respect to t

$$\frac{d}{dt}DF_t(x_0) = Df[F_t(x_0)]DF_t(x_0)$$

Which is the variational equation corresponding to the differential equation that represents the system, the part $DF_t(x_0)$ corresponds to the Jacobian of time t map evaluated in the initial condition v, the first part corresponds to the partial derivatives of the differential equation evaluated in the solution so:

$$J_t = DF_t(x_0)$$

$$A(t) = Df[F_t(x_0)]$$

Then the derivative is rewritten as $\dot{J}_t = A(t)J_t$. The Jacobian J_t, when evaluated at the initial condition v, maps small variations tangent to the trajectory at the initial time to small variations to the trajectory at time T, this is:

$$DF_t(x_0) f(x_0) = f[F_t(t)]$$

Where $F_t(x_0)$ is the trajectory at time T, the equations that determine the system f also give the direction of the trajectory at all time, the derivative of this last equation is:

$$\frac{d}{dt} DF_{Tt}(x_0) f(x_0) = Df[F_t(x_0)] DF_t(x_0) f(x_0)$$

Doing a change of variable $\delta = DF_t(x_0) f(x_0)$ satisfies

$$\dot{\delta} = Df[F_t(x_0)]\delta$$

$$\delta(0) = f(x_0)$$

Also the derivative with respect of time of the equation is

$$\frac{d^2}{dt^2} F_t(x_0) = Df[F_t(x_0)] \frac{d}{dt} F_t(x_0)$$

$$\frac{d}{dt} f[F_t(x_0)] = Df[F_t(x_0)] f[F_t(x_0)]$$

Where $f[F_t(x_0)]=f(x_0)$ at $t = 0$. An important consequence of this is that a bounded trajectory either has one Lyapunov exponent equal to zero or an equilibrium in its limit set, if this does not happen then $0 < b < |f[F_t(x_0)]| < B$ for all t with the positive bound b and B, if $\alpha(n)$ is the expansion in the direction of $f(v)$ after n time units then

$$0 \leq \lim_{n \to \infty} \frac{1}{n} \ln b \leq \lim_{n \to \infty} \frac{1}{n} \ln \alpha(n) \leq \lim_{n \to \infty} \frac{1}{n} \ln B \leq 0$$

Hence the Lyapunov exponent in the direction tangent of the trajectory is zero. Defining $\Delta(t) = \det J_t$ where $\dot{J}_t = A(t)J_t$ then:

$$\Delta t' = T_r[A(t)]\Delta_t$$

$$\Delta_0 = \det J_0 = 1$$

So

$$\det J_t = \exp\left[\int_0^t T_r A\,(t)\,dt\right]$$

This results makes possible to say that a system is dissipative if its time T map decreases for all T>0, a trace $T_r\,[A\,(t)] < 0$ implies that the system is dissipative.

Let $F_t\,(x_0)$ be a solution for $\dot{x} = f\,(x_0)$, the orbit is chaotic if:

1. the trajectory $F_t\,(x_0)\,,\,t \geq 0$ is bounded.
2. the trajectory $F_t\,(x_0)$ has at least a positive Lyapunov exponent

Example 2.1 Consider the Lorenz system

$$\dot{x} = -\alpha\,(x - y)$$

$$\dot{y} = \beta x - xz - y$$

$$\dot{z} = xy - bz$$

The partial derivatives of the differential equation are:

$$A\,(t) = \begin{bmatrix} -\alpha & \alpha & 0 \\ \beta - z & -1 & -x \\ y & x & -b \end{bmatrix}$$

Then $\Delta\,(t) = \exp\left[\int_0^t (-\alpha - 1 - b)\,dt\right] = e^{-(\alpha+1+b)t}$, depending on the selection of values the Lorenz system can be made chaotic or not

2.2   Stability

Secure communications based on control theory need systems with bounded dynamics, this means that it is absolutely necessary that those systems are stable, an unstable system has trajectories that never stop growing, making them useless for encryption purposes.

Various portions of the communication system need an stability analysis, for example if synchronization or observers are involved it is necessary to show that the states of the slave or observer converge to the ones of the master system i.e. the error is asymptotically stable. The most helpful method to determine the stability of the system is based on the works of Alexander M. Lyapunov named the general problem of motion stability, from this work spawn two methods to determine the stability of a system: The linearization method and the Lyapunov's direct method. The linearization method obtains information about local stability of an equilibrium point from the stability of a linear approximation of the system, the direct method determines the stability by using an energy like scalar function and is not locally limited. The linearization method is most common in stability analysis of linear systems, but when dealing with nonlinear systems, the second method is the most successful.

2.2.1 Nonlinear Systems

Dynamic systems are represented usually by a set of differential equations given by:

$$\dot{x} = f(x, t)$$

With the nonlinear vector function $f(x, t)$. A nonlinear system can be classified as autonomous and non autonomous, an autonomous systems states does not depend on the time, its equation is written as:

$$\dot{x} = f(x)$$

A non autonomous systems states depend on the time, a autonomous systems trajectories are not affected by the initial time but non autonomous trajectories are dependent on the initial time.

When the trajectories of a system converge to a point it is called an equilibrium point or equilibrium state:

Definition 2.2 A state x_e is an equilibrium point of a system $\dot{x}(t) = f(x)$ if the state $x(t)$ retains the value of x_e once it has reach that value, since the equilibrium point satisfies $0 = (x_e)$, this last equation makes possible to find equilibrium points by solving it.

Example 2.2 Find the equilibrium points of the unforced Duffing oscillator $\ddot{x} + d\dot{x} + bx + ax^3 = 0$

By doing the change of variable $\dot{x} = y$ the oscillator's space state representation is :

$$\begin{bmatrix} \dot{x} \\ \dot{y} \end{bmatrix} = \begin{bmatrix} 0 & 1 \\ -b - ax^2 & -d \end{bmatrix} \begin{bmatrix} x \\ y \end{bmatrix}$$

Fig. 2.1 State trajectories with different types of stability. Notice how the asymptotic stability implies convergence to zero

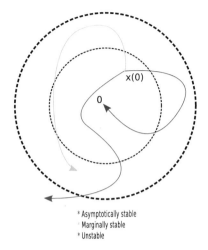

* Asymptotically stable
  Marginally stable
* Unstable

Then

$$\begin{bmatrix} 0 \\ 0 \end{bmatrix} = \begin{bmatrix} 0 & 1 \\ -b-ax^2 & -d \end{bmatrix} \begin{bmatrix} x \\ y \end{bmatrix}$$

When solving the equation it is possible to conclude that the equilibrium points are $x = \pm\sqrt{-b/a}, y = 0$.

Trajectories that enter equilibrium points usually remain within the equilibrium point, some equilibrium points have basins of attraction, these are regions that surround the equilibrium point, when a trajectory enters the basin of attraction it converges to the attraction point or remains in an orbit around it. Nonlinear systems have a variety of behaviors that may seem different than asymptotic stability and yet, they are stable (see Fig. 2.1). In order to more accurately describe the stability of nonlinear systems a formal definition of stability is given next:

Definition 2.3 The state $x = 0$ is said to be stable in the sense of Lyapunov or stable if for every number $R > 0$, $R \in \mathbb{R}$, there exist a $r > 0$, $r \in \mathbb{R}$, such that if $\|x(0)\| < r$, then for every $t \geq 0$, $\|x(t)\| < R$. If the condition is not satisfied the systems is unstable.

The definition means that the trajectory of the system remains arbitrarily close to the origin if its initial conditions are sufficiently close to the origin, this is, there is a ball of radius r such that if the initial condition of the states start within that ball, the state remains contained within a ball of radius R for all time $t \geq 0$, the system is unstable if there is at least one ball of radius R that for any $r > 0$ the trajectory does not remains contained within the ball of radius R even if the initial conditions is within the ball.

Many control systems require that a state moves to a desired equilibrium point and remain there, regardless of outside actions over the system, the next definition refers to this concept:

Definition 2.4 An stable equilibrium state $x(t) = 0$ is said to be asymptotically stable if there is $r > 0$, $r \in \mathbb{R}$ such that $\|x(0)\| < r$ and it makes the state $x(t) \rightarrow 0$ when $t \rightarrow \infty$.

Definition 2.5 An stable equilibrium state $x(t) = 0$ that is not asymptotically stable is called marginally stable.

The definition states that any state that nears to 0 converges to 0 as time passes, it will come infinitely close to 0 and reach it when $t \rightarrow \infty$. this concept of stability only states that the state will reach the desired equilibrium point after infinite time, another concept of stability considers how fast a state converges to the equilibrium point:

Definition 2.6 A stable equilibrium state $x(t) = 0$ is said to be exponentially stable if there are numbers $a > 0$, $a \in \mathbb{R}$ and $\lambda > 0$, $\lambda \in \mathbb{R}$ that satisfy:

$$\|x(t)\| \leq a \|x(0)\| e^{-\lambda t}, \ \forall t > 0$$

The exponential convergence encompasses asymptotic stability and allows to bound the states during their whole trajectory, the following image illustrates the different types of stability:

Definition 2.7 If an exponentially or asymptotically stable equilibrium point remains as it is for any initial condition, the equilibrium point is said to be globally exponentially stable or globally asymptotically stable, otherwise the system is locally stable only.

2.2.2 Stability and Linearization

Lyapunov stability has to main methods, the first one is about local stability of nonlinear systems, it borrows from the idea that a nonlinear system should behave very similarly to a linearized approximation of it when the initial conditions are not to far from the stable equilibrium point. The linearization is a simple process, consider the nonlinear system:

$$\dot{x} = f(x)$$

The Taylor expansion of the system is

$$\dot{x} = \left(\frac{\partial f}{\partial x}\right)_{(x=ep)} x + H_o(x)$$

where $H_o(x)$ are the higher order terms and $\left(\frac{\partial f}{\partial x}\right)$ is the Jacobian matrix defined by:

$$J = \left[\frac{\partial f}{\partial x_1} \cdots \frac{\partial f}{\partial x_n}\right] = \begin{bmatrix} \frac{\partial f_1}{\partial x_1} & \cdots & \frac{\partial f_1}{\partial x_n} \\ \vdots & \ddots & \vdots \\ \frac{\partial f_m}{\partial x_1} & \cdots & \frac{\partial f_m}{\partial x_n} \end{bmatrix}$$

Evaluated in the equilibrium point ep. Thus the linear approximation of the system can be written as:

$$\dot{x} = Ax, \quad A = J$$

The method to obtain a linearization of the system for a non autonomous system is similar, having the system with input u:

$$\dot{x} = f(x, u)$$

The Taylor expansion for the non autonomous system is:

$$\dot{x} = \left(\frac{\partial f}{\partial x}\right)_{(x=ep, u=0)} x + \left(\frac{\partial f}{\partial u}\right)_{(x=ep, u=0)} u + H_o(x, u)$$

With $H_o(x, u)$ being the higher order terms, then making the change of variable for the Jacobian matrices $A = \left(\frac{\partial f}{\partial x}\right)$ and $B = \left(\frac{\partial f}{\partial u}\right)$ gives the linearization of the non autonomous nonlinear system:

$$\dot{x} = Ax + Bu$$

The Lyapunov linearization method allows to conclude about the stability of the nonlinear system that:

- The equilibrium point is asymptotically stable if the eigenvalues of the matrix A of the linearized system are in the left half of the complex plane.
- The equilibrium point is unstable if the linearized system is unstable
- If at least one of the eigenvalues of the matrix A is in the vertical axis of the complex plane, it is impossible to draw a conclusion about the stability of the system with the linearization.

Example 2.3 The unforced Duffing equation has the next space state representation:

$$\dot{x} = y$$

$$\dot{y} = -ay - bx - cx^3$$

where $a, b, c \in \mathbb{R}$. The partial derivatives for each equation are:

$$\frac{\partial f_1}{\partial x} = 0$$

$$\frac{\partial f_1}{\partial y} = 1$$

$$\frac{\partial f_2}{\partial x} = -bx - 3cx^2$$

$$\frac{\partial f_2}{\partial y} = -ay$$

The terms are then arranged in a more convenient form:

$$\frac{\partial f}{\partial x, y} = \begin{bmatrix} 0 & 1 \\ -bx - 3cx^2 & -ay \end{bmatrix}$$

It is desired to know the stability around the initial condition. $x_0 = 1$, $y_0 = 1$ thus substituting these values in the previous equation leads to:

$$\frac{\partial f}{\partial x_0 = 0,\ y_0 = 0} = \begin{bmatrix} 0 & 1 \\ -b(1) - 3c(1)^2 & -a(1) \end{bmatrix}$$

then the linearized system is

$$\begin{bmatrix} \dot{x} \\ \dot{y} \end{bmatrix} = \begin{bmatrix} 0 & 1 \\ -b - 3c & -a \end{bmatrix} \begin{bmatrix} x \\ y \end{bmatrix}$$

The characteristic polynomial of the linearized system is $s^2 + as + (b + 3c) = 0$, choosing $a = 5$, $b = 1$ and $c = 1$ gives the eigenvalues $s_1 = -1$, $s_1 = -4$ making the linearized system stable. The states of both, the linearized and nonlinear Duffing unforced equation, can be observed in Fig. 2.2. Notice that, although the states are not exactly the same, these show a very similar behavior and over time they converge to the equilibrium point. Depending on how dominant is the nonlinear part of the system, the linearization resembles more or less the nonlinear system.

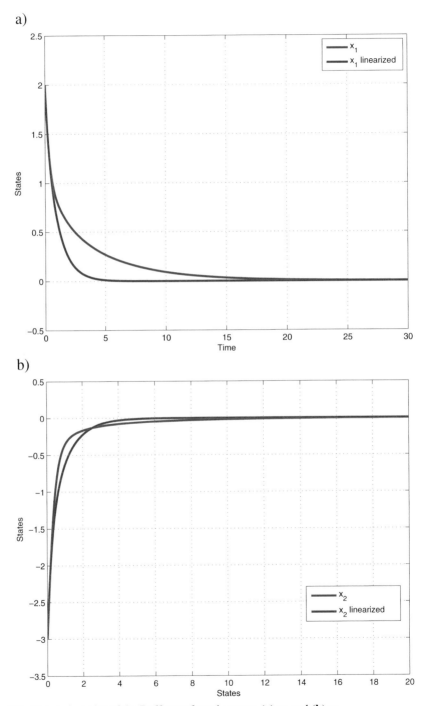

Fig. 2.2 State trajectories of the Duffing unforced system, (**a**) x_1 and (**b**) x_2

Example 2.4 Another interesting chaotic dynamic system is the Van Der Pol oscillator. It also has two states and is represented by the equation:

$$\dot{x} = y$$
$$\dot{y} = \mu \left(1 - x^2\right) y - x$$

Where the real value $\mu > 0$ controls the nonlinear damping. This system has the following partial derivatives:

$$\frac{\partial f_1}{\partial x} = 0$$

$$\frac{\partial f_1}{\partial y} = 1$$

$$\frac{\partial f_2}{\partial x} = -2\mu x y - 1$$

$$\frac{\partial f_2}{\partial y} = -\mu x^2 + \mu$$

Then, rewriting the equation:

$$\frac{\partial f}{\partial x, y} = \begin{bmatrix} 0 & 1 \\ -2\mu x y - 1 & -\mu x^2 + \mu \end{bmatrix}$$

The evaluation of the stability around the point $x_0 = 0.5$, $y_0 = 0$ is done by evaluating the partial derivatives around this point:

$$\frac{\partial f}{\partial x, y} = \begin{bmatrix} 0 & 1 \\ -2\mu\,(0.5)\,(0) - 1 & -\mu\,(0.5)^2 + \mu \end{bmatrix}$$

Thus the linearized system is:

$$\begin{bmatrix} \dot{x} \\ \dot{y} \end{bmatrix} = \begin{bmatrix} 0 & 1 \\ -1 & 0.75\mu \end{bmatrix} \begin{bmatrix} x \\ y \end{bmatrix}$$

The Characteristic polynomial of the linearized system is $s^2 - 0.75\,\mu s + 1 = 0$, being the nonlinear system chaotic. When the parameters are selected to produce chaos, the dynamic is bounded, but the linearized system diverges as it is an unstable linear system. By selecting $\mu = 5$, the linearized system has 0.2889 and 3.4611 as eigenvalues. The resulting dynamic can be observed in Fig. 2.3

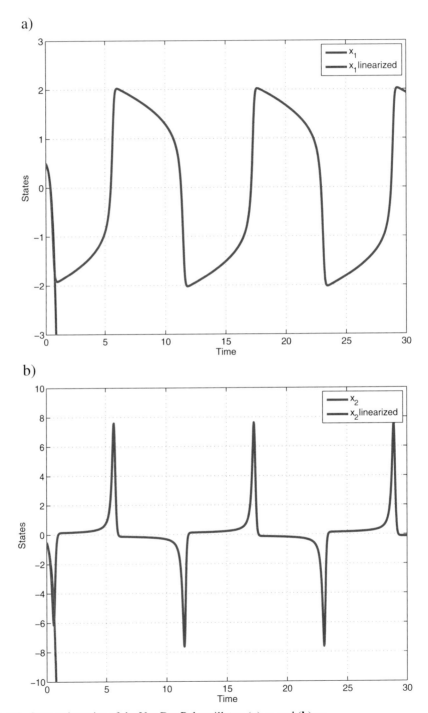

Fig. 2.3 State trajectories of the Van Der Pol oscillator, (**a**) x_1 and (**b**) x_2

As expected, the chaotic system remains bounded and the states of the linearized system diverge. If the parameter is selected such that the linearized system is stable, both systems behave similarly, however, chaotic behavior is lost as the system is now asymptotically stable.

This example shows a significant difference in the stability of nonlinear systems compared to the linearized equivalent. This limitation makes this approach less helpful for the stability analysis of chaotic systems.

Example 2.5 The Rössler chaotic oscillator is defined by the set of equations:

$$\dot{x} = -y - z$$
$$\dot{y} = x + ay$$
$$\dot{z} = b + z(x - c)$$

The oscillator uses the parameters $a, b, c \in \mathbb{R}$. Changing these values affects the shape of the attractor and the chaotic behavior. Then, the partial derivatives of the system are:

$$\frac{\partial f_1}{\partial x} = 0$$

$$\frac{\partial f_1}{\partial y} = -1$$

$$\frac{\partial f_1}{\partial z} = -1$$

$$\frac{\partial f_2}{\partial x} = 1$$

$$\frac{\partial f_2}{\partial y} = a$$

$$\frac{\partial f_2}{\partial z} = 0$$

$$\frac{\partial f_3}{\partial x} = z$$

$$\frac{\partial f_3}{\partial y} = 0$$

$$\frac{\partial f_3}{\partial z} = x - c$$

these results are arranged in a convenient way:

$$\frac{\partial f}{\partial x, y, z} = \begin{bmatrix} 0 & -1 & -1 \\ 1 & a & 0 \\ z & 0 & x-c \end{bmatrix}$$

It is desired to know the stability around the origin $x_0 = 0$, $y_0 = 0$, $z_0 = 0$, evaluating in the origin yields the linearized system:

$$\begin{bmatrix} \dot{x} \\ \dot{y} \\ \dot{z} \end{bmatrix} = \begin{bmatrix} 0 & -1 & -1 \\ 1 & a & 0 \\ 0 & 0 & -c \end{bmatrix} \begin{bmatrix} x \\ y \\ z \end{bmatrix}$$

With $s^3 + (c-a)s^2 + (1-ac)s + c = 0$ as characteristic polynomial. This can be further factorized into the expression $(s+c)(s^2 - as + 1)$. Then, if the values $a = -2$, $b = 0.01$, $c = 3$ are selected, the linearized system has -1, -1 and -3 as eigenvalues and it behaves as is shown in Fig. 2.4.

The Linearization approximately has the same behavior around the origin, but if the system is made chaotic the linearized equivalent becomes unstable.

Exercise 2.1 Obtain a conclusion about the stability of the Chua oscillator:

$$\dot{x} = a\,[y - x - g\,(x)]$$
$$\dot{y} = b\,(x - y + z)$$
$$\dot{z} = -c\,(y)$$
$$g\,(x) = d_1 x + \left(\frac{d_0 - d_1}{2}\right)(|x+1| - |x-1|)$$

By applying the linearization method, consider that the values $a, b, c, d_0, d_1 \in \mathbb{R}$ can be selected to affect the stability of the system.

Exercise 2.2 Obtain a conclusion about the stability of the pendulum given by:

$$\dot{x} = y$$
$$\dot{y} = -a \sin x - by$$

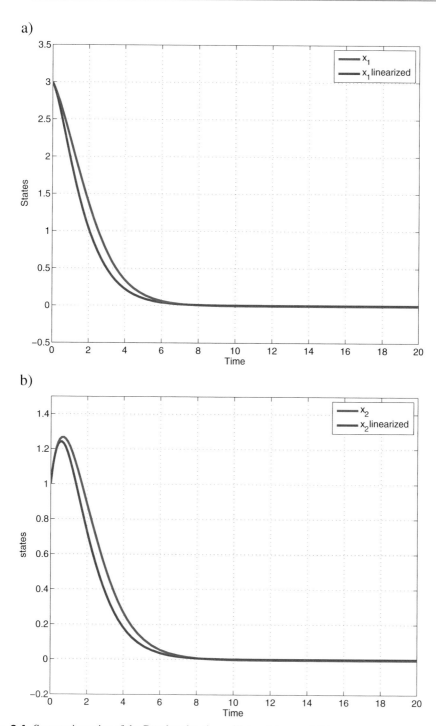

Fig. 2.4 State trajectories of the Rössler chaotic oscillator, (**a**) x_1 and (**b**) x_2

Exercise 2.3 Obtain a conclusion about the stability system:

$$\dot{x} = x \cos y$$

$$\dot{y} = x(1 + y) - x^2$$

2.2.3 Lyapunov's Direct Method

The Lyapunov's direct method takes the idea from physical system, where the system has a certain amount of energy, when the system consumes said energy after some time it depletes all of it and comes to a stop or equilibrium point. Thus, there is a similarity with the stability of a nonlinear system, where a stable state depletes its energy and falls into and equilibrium point and remains thereafter. Before advancing onto the needed stability theorems, it is necessary to provide some definitions:

Definition 2.8 A function $V(x)$ is said to be positive semidefinite if $V(x) \geq 0, \forall x \neq 0$ and $V(x) = 0, \ x = 0$.

Definition 2.9 A function $V(x)$ is said to be negative semidefinite if $V(x) \leq 0, \forall x \neq 0$ and $V(x) = 0, \ x = 0$.

Definition 2.10 A function $V(x)$ is said to be positive definite if $V(x) > 0, \forall x \neq 0$ and $V(x) = 0, \ x = 0$.

Definition 2.11 A function $V(x)$ is said to be negative definite if $V(x) < 0, \forall x \neq 0$ and $V(x) = 0, \ x = 0$.

Theorem 2.1 *Lyapunov Stability: Let $x = 0$ be an equilibrium point for the system $\dot{x} = f(x)$ If there exists a continuously differentiable positive definite function $V(x)$ and its derivative $\dot{V}(x)$ is negative semidefinite the equilibrium point 0 is stable, if $\dot{V}(x)$ is negative definite the system is asymptotically stable.*

Proof Having $\varepsilon > 0$ and $r \in (0, \varepsilon]$ that make $B_R = \{x \in \mathbb{R}^n \mid \|x\| \leq r\}$, let $a = \min_{\|x\|=r} V(x)$, $a > 0$ and $b \in (0, \alpha)$, so that $B_r = \{x \in B_R \mid V(x) \leq b\}$.

Since the derivative $\dot{V}[x(t)] \leq 0$ implies that $V[x(0)] \leq V(0) \leq b, \ \forall t > 0$ because B_r is a compact set, since $V(x)$ is continuous and $V(0) = 0$ there is a positive real number $d > 0$ that fulfills

$$\|x\| \leq d$$

and

$$V(x) < b$$

Then $x(0) \in B_d$, $x(0) \in B_R$, $x(t) \in B_R$ and $x(t) \in B_r$, hence $\|x(0)\| < d$ implies that $x(t) < r \le \varepsilon$ for all $t \ge 0$, this last part shows that the equilibrium point is stable. Since $V(x)$ is monotonically decreasing and bounded by zero, $V[x(t)] \to c \ge 0$ as $t \to \infty$, suppose that $c > 0$, the continuity of $V(x)$ implies that there is a $\delta > 0$ such that $B_\delta \in B_c$, the limit $V[x(t)] \to c > 0$ implies that the trajectory $x(t)$ is outside the ball B_δ, let $-y = max_{\delta \le \|x\| \le R} \dot{V}(x)$ caused by the continuous function $\dot{V}(x)$ having a maximum over the compact set $\delta \le \|x\| \le R$ and since $-y < 0$ it follows that

$$V[x(t)] = V[x(0)] + \int_0^t \dot{V}[x(\tau)]\,d\tau \le V[x(0)] - yt$$

Since the right side turns into a negative number as time passes, the inequality contradicts the assumption that $c > 0$ showing that $c = 0$ and that the equilibrium is asymptotically stable.   □

Example 2.6 It was shown that the linearized Duffing oscillator has a stable dynamic, for convenience it is written as follows:

$$\begin{bmatrix} \dot{x}_1 \\ \dot{x}_2 \end{bmatrix} = \begin{bmatrix} 0 & 1 \\ -b - 3c & -a \end{bmatrix} \begin{bmatrix} x_1 \\ x_2 \end{bmatrix}$$

Then we propose to use the Lyapunov candidate function $V(x) = x^T x$ to show it is indeed asymptotically stable, the derivative is:

$$\dot{V}(x) = x^T \dot{x} + \dot{x}^T x$$

$$= x^T A x + x^T A^T x$$

$$= \begin{bmatrix} x_1 & x_2 \end{bmatrix} \begin{bmatrix} 0 & 1 \\ -b - 3c & -a \end{bmatrix} \begin{bmatrix} x_1 \\ x_2 \end{bmatrix} + \begin{bmatrix} x_1 & x_2 \end{bmatrix} \begin{bmatrix} 0 & -b - 3c \\ 1 & -a \end{bmatrix} \begin{bmatrix} x_1 \\ x_2 \end{bmatrix}$$

$$= \begin{bmatrix} x_1 & x_2 \end{bmatrix} \begin{bmatrix} x_2 \\ -ax_2 - (b + 3c)x_1 \end{bmatrix} + \begin{bmatrix} x_1 & x_2 \end{bmatrix} \begin{bmatrix} -(b + 3c)x_2 \\ x_1 - ax_2 \end{bmatrix}$$

$$= [x_2(x_1 - ax_2) - x_1 x_2(b + 3c)] + [x_1 x_2 - x_2(ax_2 + x_1(b + 3c))]$$

$$= -2ax_2^2 - x_1 x_2(2b + 6c - 2)$$

Choosing the values $a = 5, b = 1, c = 0$ affects the stability of the system and causes it to have eigenvalues $\lambda_1 = -0.2087, \lambda_2 = -4.7913$ then:

$$\dot{V}(x) = -10x_2^2 - x_1x_2\,(2 + 0 - 2)$$
$$= -10x_2^2$$

Thus the system is stable in the sense of Lyapunov:

$$\dot{V}(x) \le 0$$

But knowing that the system is linear, then it is possible to conclude that it is also asymptotically stable, as it is shown in Fig. 2.5.

Both states converge asymptotically to the origin as indicated by their chosen stable poles.

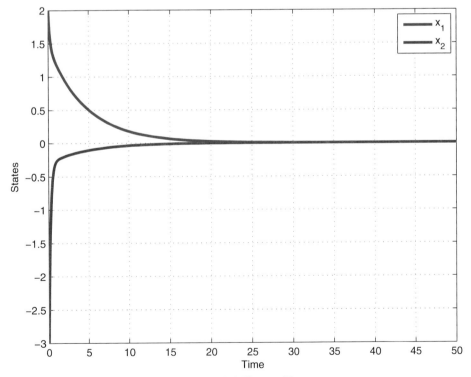

Fig. 2.5 State trajectories of a stable linearized duffing oscillator

Example 2.7 Since the Lyapunov's direct stability is meant to analyze non linear systems stability, let us find out about the stability of the nonlinear forced Duffing oscillator, as previously, for convenience, it is rewritten:

$$\dot{x}_1 = x_2$$
$$\dot{x}_2 = -ax_2 - bx_1 - cx_1^3 + \gamma \cos(\omega t)$$

The same Lyapunov candidate function is useful in this case $V(x) = x^T x$, the derivative is

$$V(x) = x^T \dot{x} + \dot{x}^T x$$

$$= \begin{bmatrix} x_1 & x_2 \end{bmatrix} \begin{bmatrix} x_2 \\ -ax_2 - bx_1 - cx_1^3 \end{bmatrix} + \begin{bmatrix} x_2 & -ax_2 - bx_1 - cx_1^3 \end{bmatrix} \begin{bmatrix} x_1 \\ x_2 \end{bmatrix}$$

$$= 2x_1 x_2 - 2x_2 \left(cx_1^3 + bx_1 + ax_2 \right)$$

The nonlinear nature of the system makes impossible to assure that the derivative be $V(x) < 0$ nor $V(x) \leq 0$ regardless of the choosing of the values a,b and c, this situation was expected, as chaotic systems are not asymptotically stable nor stable, they are only bounded, this mean that they do not explode like an unstable linear system would, but also do not converge to an equilibrium point either, instead they rather "oscillate" around the equilibrium point. This phenomenon can be better appreciated in Fig. 2.6.

Exercise 2.4 Determine the stability conditions for the forced Duffing oscillator:

$$\dot{x} = y$$
$$\dot{y} = -ay - bx - cx^3 + d \cos \omega t$$

Exercise 2.5 Give a conclusion about the stability of the Van Der Pol equation:

$$\dot{x} = y$$
$$\dot{y} = \mu \left(1 - x^2 \right) y - x$$

Exercise 2.6 Give a conclusion about the stability of the Chua equation:

$$\dot{x} = a[y - x - g(x)]$$
$$\dot{y} = b(x - y + z)$$

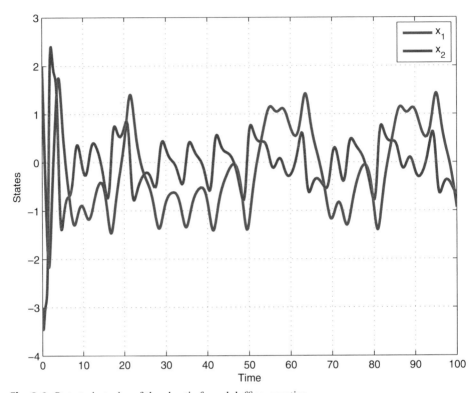

Fig. 2.6 State trajectories of the chaotic forced duffing equation

$$\dot{z} = -c\,(y)$$

$$g\,(x) = d_1 x + \left(\frac{d_0 - d_1}{2}\right)(|x+1| - |x-1|)$$

Chaotic systems fulfill several instability theorems based on the Lyapunov stability method, these could be useful when dealing with chaotic systems.

Theorem 2.2 *First instability theorem: If within a region Ω surrounding the origin there is a Lyapunov function $V\,(x,t)$ such that $V\,(0,t) = 0$, $\forall t \geq t_0$, $V\,(x,t_0) > 0$ and $\dot{V}\,(x,t)$ is positive definite, the equilibrium point 0 is unstable at $t = t_0$.*

Theorem 2.3 *Second instability theorem: If within a region Ω surrounding the origin there is a Lyapunov function $V\,(x,t)$ which fulfills that $V\,(0,t_0) = 0$ and $V\,(x,t_0)$ are positive definite near the origin and the derivative $\dot{V}\,(x,t) - \lambda V\,(x,t) \geq 0$ where $\lambda > 0$ makes the equilibrium be unstable.*

Theorem 2.4 *Third instability theorem: If within a region Ω surrounding the origin there is a Lyapunov function $V(x,t)$ that has first partial derivatives that decrease and $V(x,t)$ and $\dot{V}(x,t)$ are positive definite, the origin is a boundary of the subregion of Ω called Ω_s and at the boundary points of Ω_s $V(x,t) = 0$, causes the equilibrium point to be unstable.*

Example 2.8 Consider the previous stability result on the unforced Duffing oscillator

$$\dot{x}_1 = x_2$$
$$\dot{x}_2 = -ax_2 - bx_1 - cx_1^3$$

Where the Lyapunov candidate function $V(x) = x^T x$ yields the derivative

$$\dot{V}(x) = 2x_1 x_2 - 2x_2 \left(cx_1^3 + bx_1 + ax_2 \right)$$
$$= 2x_1 x_2 - 2bx_1 x_2 - 2cx_1^3 x_2 - 2ax_2^2$$

Choosing $a = -1$, $b = -1$ and $c = 0$ lead to

$$\dot{V}(x) = 2ax_2^2$$
$$\dot{V}(x) < 2ax_2^2 + 1$$

Making the system unstable as stated in the first instability theorem.

Exercise 2.7 Determine which set of parameters make the Rössler equation unstable as stated in the second instability theorem;

$$\dot{x} = -y - z$$
$$\dot{y} = x + ay$$
$$\dot{z} = b + z(x - c)$$

Exercise 2.8 Determine the set of parameters that make the Chua oscillator unstable according to the third instability theorem:

$$\dot{x} = a[y - x - g(x)]$$
$$\dot{y} = b(x - y + z)$$
$$\dot{z} = -c(y)$$
$$g(x) = d_1 x + \left(\frac{d_0 - d_1}{2} \right)(|x + 1| - |x - 1|)$$

Having covered the most important results about stability that are to be relied upon in the rest of the book, the next step is to introduce state observers, that are one of the most important part in secure communications based on non linear control theory.

2.3   State Observers

In many cases, dynamical systems do not have all states available to be directly measured or the sensors required for their measurement are not available, then, it is necessary to estimate the state variables that are not known, the process of estimating unknown state variables is named observation and the device that computes the estimate is called a state observer or simply observer. A state observer estimates the unknown state variables based on the measure of the state variables that are accessible for the sensors of the system.

A very simple, yet effective solution to this problem is the Luenberger observer, named after David G. Luenberger, it is most known and used for linear system's observation, but it is also capable of working well with non linear chaotic systems, in the next section the details of its workings are given.

2.3.1   Luenberger Observer

The Luenberger state observer is a full dimensional state estimator, this means that it makes a reconstruction of all the states of the desired system. The observer is basically a copy of the system with a set of closed loop correction parameters that make the estimation error asymptotically stable [4].

Consider the non linear autonomous system:

$$\dot{x} = Ax + f(x)$$
$$y = Cx$$

where the matrix A is the linear part of the system and the vector $f(x)$ contains the nonlinear parts of each state, the Luenberger state observer for the system is:

$$\dot{\hat{x}} = A\hat{x} + f(\hat{x}) + KCe$$
$$\hat{y} = C\hat{x}$$

Where $\hat{x}$ is the estimate of the state, K is a gain vector and $e = x - \hat{x}$ is the estimation error. Using the Lyapunov method for stability analysis gives good results and produces an easy way to compute the values of K, the derivative of the error is

$$\dot{e} = Ax + f(x) - A\hat{x} - f(\hat{x}) - KCe$$
$$= Ae + f(x) - f(\hat{x}) - KCe$$

Making the change of variable $\phi(e) = f(x) - f(\hat{x})$ generates the error dynamic equation:

$$\dot{e} = Ae + \phi(e) - KCe$$

The proof of stability is based on the following assumptions:

Assumption 2.1 *There is solution to the linear matrix inequality $A^T P + PA + 2\alpha P + \beta I < Q$ with the positive definite symmetric matrices $P = P^T > 0$, $Q = Q^T > 0$ and the real numbers $\alpha > 0, \beta > 0$.*

Assumption 2.2 *The nonlinear part of the equation complies with the equation $x^T P \phi(x) \le 2\alpha x^T Px + x^T x$.*

The next Lyapunov equation and its derivative are used for the proof of stability:

$$V = e^T Pe$$
$$\dot{V} = \dot{e}^T Pe + e^T P\dot{e}$$

Proof The derivative is rewritten:

$$\dot{V} = [Ae + \phi(e) - KCe]^T Pe + e^T P[Ae + \phi(e) - KCe]$$
$$= e^T A^T Pe + \phi(e)^T Pe - (KCe)^T Pe + e^T PAe + e^T P\phi(e) - e^T PKCe$$
$$= e^T A^T Pe + \phi(e)^T Pe + e^T PAe + e^T P\phi(e) - (KCe)^T Pe - e^T PKCe$$

By assumption 2.2

$$\dot{V} \le e^T A^T Pe + e^T PAe + 2\alpha e^T Pe + e^T e - (KCe)^T Pe - e^T PKCe$$
$$\le e^T \left[A^T P + PA + 2\alpha P + I \right] e - 2e^T PKCe$$

If assumption 2.1 is met

$$\dot{V} < e^T [Q - PKC]e$$

The gain vector K is then selected to make the matrix $Q - PKC$ definite negative thus making the error asymptotically stable in the origin, allowing the observer to reconstruct the states.

The result above implies that the error is asymptotically stable around the origin, showing that the states of the observer converge to the states of the targeted system, then it is safe to assume that after some time, the error is negligible and the observer states are almost the same as the target system. the next examples show how the states converge and the asymptotic stability around the origin that the error possess.

Example 2.9 The forced Duffing oscillator is represented by the state equation:

$$\dot{x}_1 = x_2$$

$$\dot{x}_2 = -ax_2 - bx_1 - cx_1^3 + d\cos\omega t$$

$$y = x_1$$

The matrices $A = \begin{bmatrix} 0 & 1 \\ -b & -a \end{bmatrix}$, $f(x) = \begin{bmatrix} 0 \\ -cx_1^3 + d\cos\omega t \end{bmatrix}$, $x = \begin{bmatrix} x_1 \\ x_2 \end{bmatrix}$ and $C = \begin{bmatrix} 1 & 0 \end{bmatrix}$ allow to express the system as:

$$\dot{x} = Ax + f(x)$$

$$y = Cx$$

A Luenberger observer for the system is

$$\dot{\hat{x}} = A\hat{x} + f(\hat{x}) + KCe$$

$$\hat{y} = C\hat{x}$$

Where $e = \begin{bmatrix} x_1 - \hat{x}_1 \\ x_2 - \hat{x}_2 \end{bmatrix}$ and $K = \begin{bmatrix} k_1 \\ k_2 \end{bmatrix}$. The obtained results by using the values $a = 0.2, b = -1, c = 1, d = 0.3, \omega = 1, k_1 = 5$ and $k_2 = 10$ are shown in Figs. 2.7 and 2.8.

If the gains are properly selected the error is asymptotically stable as the graphic shows, the attractors converge and it is possible to see that the initial condition of the observer is different than the Duffing oscillator, and after a short while it falls into the equilibrium point 0 and remains thereafter.

a)

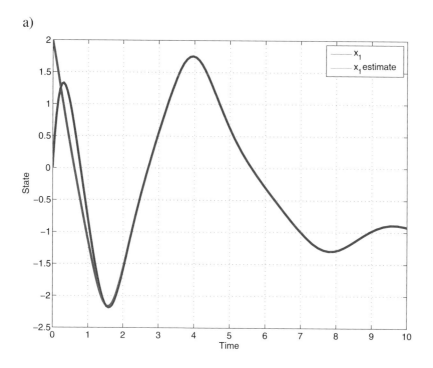

b)

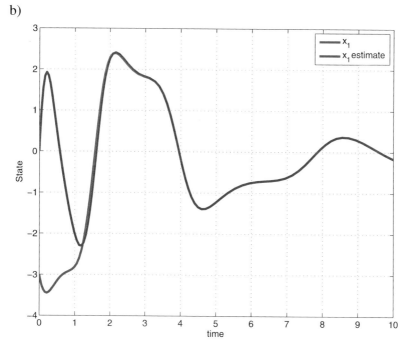

Fig. 2.7 Forced duffing oscillator, state trajectories and estimates, (**a**) x_1 and (**b**) x_2

a)

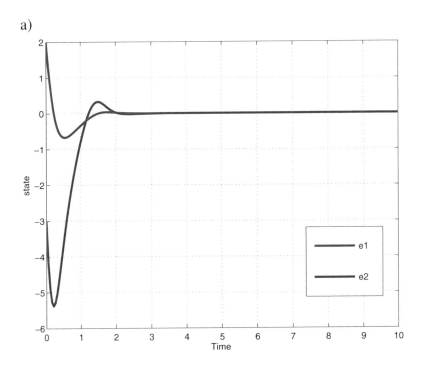

b)

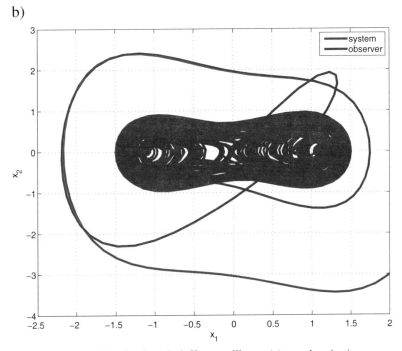

Fig. 2.8 State estimation for the forced duffing oscillator, (**a**) synchronization error and (**b**) convergence of both trajectories (oscillator and observer)

Example 2.10 The Van Der Pol dynamic equation is written as follows:

$$\dot{x}_1 = x_2$$
$$\dot{x}_2 = \mu\left(1 - x_1^2\right)x_2 - x_1$$
$$y = x_1$$

A Luenberger observer for the system has the following state equations:

$$\dot{\hat{x}}_1 = \hat{x}_2 + k_1 e$$
$$\dot{\hat{x}}_2 = \mu\left(1 - \hat{x}_1^2\right)\hat{x}_2 - \hat{x}_1 + k_2 e$$
$$y = \hat{x}_1$$

The observer can be expressed by the matrices:

$$\begin{bmatrix} \dot{\hat{x}}_1 \\ \dot{\hat{x}}_2 \end{bmatrix} = \begin{bmatrix} 0 & 1 \\ -1 & -\mu \end{bmatrix}\begin{bmatrix} \hat{x}_1 \\ \hat{x}_2 \end{bmatrix} + \begin{bmatrix} 0 \\ -\mu\hat{x}_1^2\hat{x}_2 \end{bmatrix} + \begin{bmatrix} k_1 \\ k_2 \end{bmatrix}\begin{bmatrix} 1 & 0 \end{bmatrix}\begin{bmatrix} \hat{x}_1 - x_1 \\ \hat{x}_2 - x_2 \end{bmatrix}$$
$$\hat{y} = \hat{x}_1$$

The Van Der Pol oscillator exhibits a bounded dynamic when $\mu = 5$ and using the gains $k_1 = 3$ and $k_2 = 6$ the observers behavior can be seen in the Figs. 2.9 and 2.10.

Again the error is asymptotically stable around the origin, both state quickly converge as does the attractor.

Example 2.11 The Rössler oscillator:

$$\dot{x}_1 = -x_2 - x_3$$
$$\dot{x}_2 = x_1 + ax_2$$
$$\dot{x}_3 = b + x_3(x_1 - c)$$
$$y = x_2$$

The values $a = 0.2$, $b = 0.2$, $c = 5.7$ give a bounded dynamic, the state estimator has gains: $k_1 = 12$, $k_2 = 5$ and $k_3 = 17$ with the next differential equation system:

$$\dot{\hat{x}}_1 = -\hat{x}_2 - \hat{x}_3 - k_1 e$$
$$\dot{\hat{x}}_2 = \hat{x}_1 + a\hat{x}_2 - k_2$$
$$\dot{\hat{x}}_3 = b + \hat{x}_3(\hat{x}_1 - c) - k_3 e$$
$$y = \hat{x}_2$$

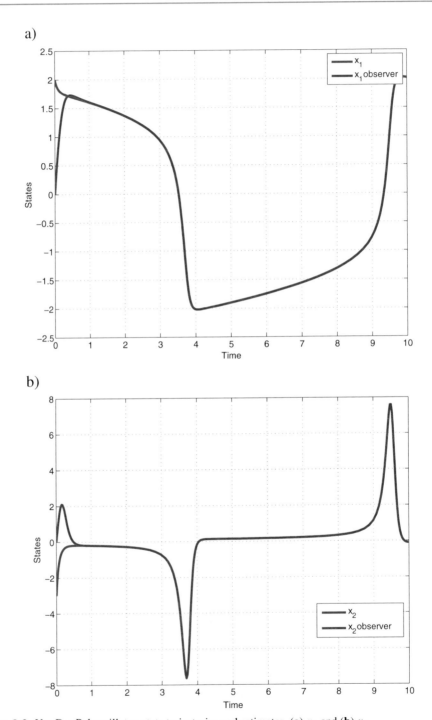

Fig. 2.9 Van Der Pol oscillator, state trajectories and estimates, (**a**) x_1 and (**b**) x_2

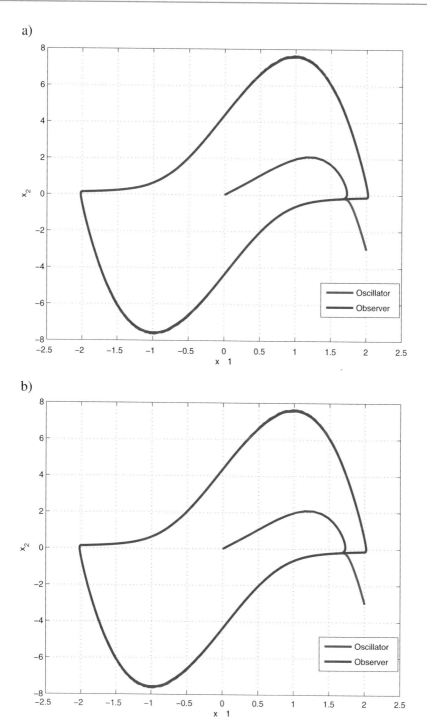

Fig. 2.10 State estimation for the Van Der Pol oscillator, (**a**) synchronization error and (**b**) convergence of both trajectories (oscillator and observer)

The resulting graphics are shown in Figs. 2.11 and 2.12

Exercise 2.9 Design a Luenberger state estimator for the Linearized Duffing equation:

$$\begin{bmatrix} \dot{x}_1 \\ \dot{x}_2 \end{bmatrix} = \begin{bmatrix} 0 & 1 \\ -b - 3c & -a \end{bmatrix} \begin{bmatrix} x_1 \\ x_2 \end{bmatrix}$$
$$y = x_1$$

Exercise 2.10 Design a Luenberger observer for the linearized Van Der Pol system:

$$\begin{bmatrix} \dot{x}_1 \\ \dot{x}_2 \end{bmatrix} = \begin{bmatrix} 0 & 1 \\ -1 & 0.75\mu \end{bmatrix} \begin{bmatrix} x_1 \\ x_2 \end{bmatrix}$$
$$y = x_1$$

Exercise 2.11 Design a Luenberger state estimator for the Chua oscillator:

$$\dot{x}_1 = a \left[x_2 - x_1 - g\left(x_1\right) \right]$$
$$\dot{x}_2 = b \left(x_1 - x_2 + x_3 \right)$$
$$\dot{x}_3 = -c \left(x_2 \right)$$
$$y = x_2$$
$$g\left(x\right) = d_1 x + \left(\frac{d_0 - d_1}{2} \right) \left(|x + 1| - |x - 1| \right)$$

2.4   Fractals and Synchronization

So far only continuous nonlinear systems have been analyzed, but there are many types of discrete chaotic systems that are also important such as Logistic maps or fractals, this last chaotic systems can comprehend real numbers or complex numbers, but all of them are self similar and have symmetry at various scales, this means that the fractal structure replicates itself at different scales i.e. it is made of the same structure at different scales and repeats itself at any scale or magnification done into the structure, this is known as self similarity, the main difference between a fractal and a non fractal object is that when the non fractal is scaled up it does not contains the original object, as an example the Fig. 2.13 it is shown a circle.

When the circle is magnified and does not contain any other circle nor it resembles the figure it was magnified from, the circle lacks of self similarity.

Fig. 2.11 Rössler oscillator, state trajectories and estimates, (**a**) x_1, (**b**) x_2 and (**c**) x_3

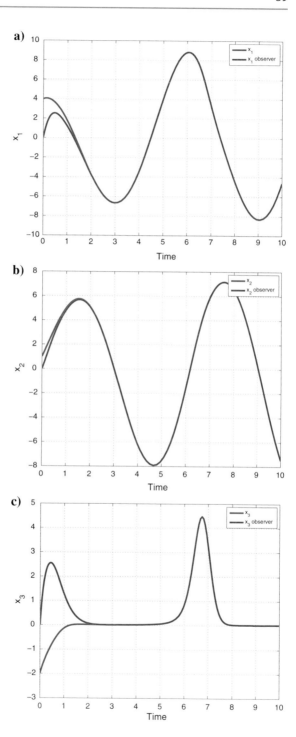

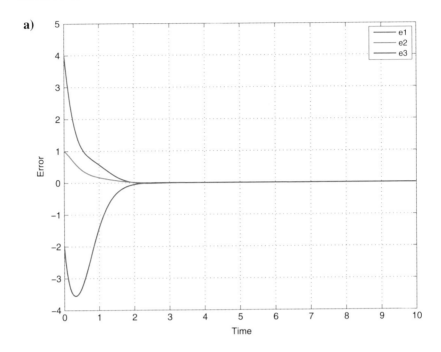

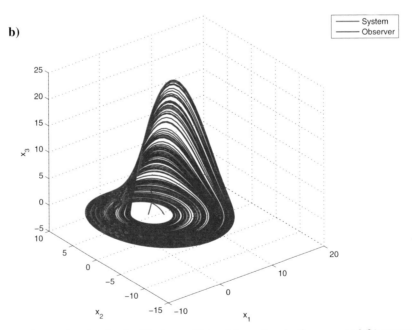

Fig. 2.12 State estimation for the Rössler oscillator, (**a**) synchronization error and (**b**) convergence of both trajectories (oscillator and observer)

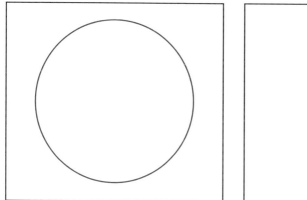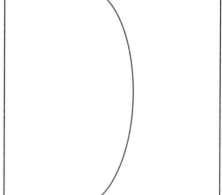

Fig. 2.13 Non autosimilar shape

The main properties of fractals are listed below [5]:

1. Fractals have self similarity, this means that if a part of the fractal is magnified it resembles the complete object.
2. The topological dimension of the fractal is always less than its fractal dimension.
3. The border of the fractal is not defined as it changes due to self similarity.
4. Fractals are represented by discrete dynamic functions or series.

The focus of this section is complex numbers fractals, in particular Mandelbrot and Julia sets, consider the quadratic map:

$$z_{n+1} = z_n^2 + c$$

Where $z_n, c \in \mathbb{C}$, if the parameter $c = 0$ the map has the fixed point $z = 0$ that has the basin of attraction $|z| < 1$, this means that function maps a point within the unitary circle to another point within the same circle but, any other point $z < 1$ that is outside of the unitary circle goes to infinity, interestingly the values that are $z = 1$ are not contained in either groups. the Mandelbrot set builds upon this idea, it represents the basin of attraction of the values $c = z_0$ that do not diverge to infinity:

Definition 2.12 The Mandelbrot set is the basin of attraction for the quadratic map $z_{n+1} = z_n^2 + c$ where $c = z_0$.

Depending on the value of c that is within the Mandelbrot set there are bounded and unbounded trajectories hence the basin of infinity is nonempty and it has a boundary, this boundary is called the Julia set:

Definition 2.13 The Julia set is the set of repelling fixed points and periodic points along with the limit points of the Mandelbrot set.

Therefore is said that Julia sets are contained within the Mandelbrot set, also consider that the Julia set also encompasses maps with different orders that 2.

Graphically these two fractals represent the initial conditions of the function that does not diverge, this could be seen as a sort of graphic representation of the behavior of each initial condition. These are drawn as a two dimension maps of the initial conditions, in one axis is the real part of the initial condition and in the other complex part, each pixel of the graphic contains a different initial condition, the set is then iterated with each pixel, after several iterations, the unbounded values are colored the same and the convergent values are assigned a different color, also different colors can be given to individual bounded values to provide more information about the convergence of the initial conditions within the fractal.

Example 2.12 The Mandelbrot set given by the equation:

$$Z_{n+1} = Z_n^2 + c, \; Z_n, c \in \mathbb{C}$$

For this example $c = z_0$, the divergent part is represented by a blue color resulting in Fig. 2.14.

The internal part of the Mandelbrot set is bounded while all the exterior is not, the bounded part is also known as the attractor.

Example 2.13 The Julia set is defined by the equation:

$$Z_{n+1} = Z_n^a + c, \; a \in \mathbb{R}, c \in \mathbb{C}$$

It is represented the same way as the Mandelbrot set, selecting the values $a = 2$ and $c = -0.61803$, the next result shown in Fig. 2.15 is obtained.

The blue colored parts diverge and the red color parts remain bounded, both graphics can also be interpreted as a visual representation of the convergence of the initial conditions of the set.

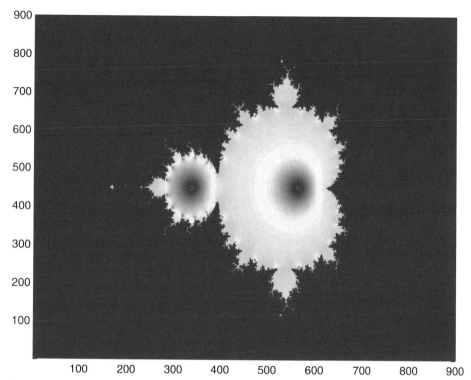

Fig. 2.14 Mandelbrot fractal

Inspired in the dynamic controllers found in [6] it is possible to design a dynamic control law that makes possible to synchronize the attractor of two fractals, and guarantee that the synchronization error decreases every iteration. Thus, consider the master system represented by:

$$M_{n+1} = f\,(M_n)$$

And the slave:

$$S_{n+1} = f\,(S_n) + u_n$$

with the dynamic control law $u_n = k\,[f\,(M_n) - f\,(S_n)]$.

Example 2.14 Synchronize the Julia set to the Mandelbrot set:

$$M_{n+1} = M_n^2 + c$$
$$S_{n+1} = S_n^a + b + u_n$$

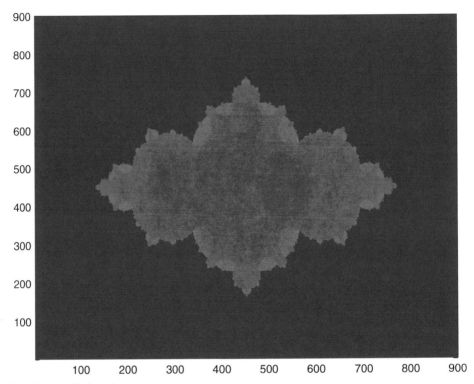

Fig. 2.15 Julia fractal

Selecting the parameters $a = 2$, $b = -0.61803$ and $c = M_0$ with the dynamic control law $u_n = 0.5 \left(M_n^2 - S_n^2 + M_0 + 0.61803 \right)$ the attractors of both sets evolve as is shown in Fig. 2.16.

The right column contains the attractors of the master at 3, 6, 9 and 25 iterations, the left column contains the attractors of the slave at 3, 6, 9 and 25 iterations, As expected the synchronization is better as the number of iterations increase, the synchronization error decreases every iteration, causing that the slave resembles the master more accurately in the second, third and fourth image rather than the first one, then it is possible to say that the convergence of the fractals is achieved, the basin of attraction of the slave closely resembles the master's basin of attraction and the bounded values converge as iterations increase.

Exercise 2.12 Draw the Julia set defined by:

$$Z_{n+1} = z_n^5 + 0.544$$

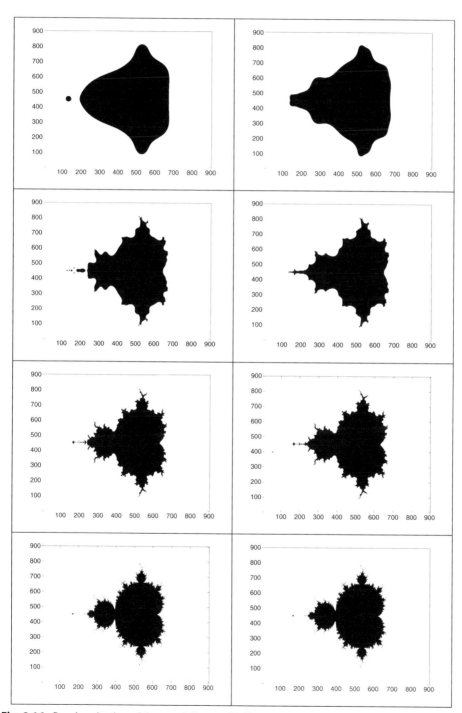

Fig. 2.16 Synchronization of the mandelbrot and julia sets

Exercise 2.13 Use the next Julia set as master

$$M_{n+1} = M_n^3 + 0.4$$

and synchronize it to the slave Julia set:

$$S_{n+1} = S_n^4 + 0.484$$

References

1. Layek, G. C. (2015). *An introduction to dynamical systems and chaos* (vol. 449). New Delhi: Springer.
2. Argyris, J. H., Faust, G., Haase, M., & Friedrich, R. (2015). *An exploration of dynamical systems and chaos: Completely revised and enlarged* (2nd edn.). Berlin: Springer.
3. Devaney, R. L. (2018). *A first course in chaotic dynamical systems: Theory and experiment*. Boca Raton: CRC Press.
4. Zeitz, M. (1987). The extended Luenberger observer for nonlinear systems. *Systems & Control Letters, 9*(2), 149–156.
5. Barnsley, M. F. (2014). *Fractals everywhere*. Cambridge: Academic Press.
6. Martínez-Guerra, R., Martínez-Fuentes, O., & Montesinos-García, J. J. (2019). *Algebraic and differential methods for nonlinear control theory: Elements of commutative algebra and algebraic geometry*. Berlin: Springer.

Stream Cyphers and Block Cyphers

3

Abstract

This chapter is about stream cyphers and block cyphers based in synchronization using the Luenberger observer.

3.1 Message and Data Carrier Signals

The purpose of encryption is making certain information accessible to only a desired group of people by modifying the message, or "hiding" it, so it is not possible to understand it without permission, the information that is desired to be transmitted is called plaintext, cleartext or message, the process of modifying the plaintext so it is not understandable is called encryption, the resulting modified message from the encryption is called ciphertext and the process of deciphering the ciphertext is called decryption, all these processes must depend on an encryption key that is usually a large set of values which can be randomly generated [1, 2]. This relation is exemplified next:

Consider a message S with the key K, the encryption function E is:

$$E(S, K) = C$$

The encryption function yields a ciphertext C that has the same size and type of data that the plaintext, when this ciphertext and the key are the input of the decryption function D:

$$D(C, K) = S$$

The result is the plaintext S, the two functions have the property

$$D\left(E\left(S, K\right), K\right) = S$$

There are also algorithms that have different keys for encryption and decryption, they are called asymmetric or public key encryption, but the focus of this book is to be symmetric key algorithms that are the ones where the same key is used for encryption and decryption.

Messages can be any type of data; video, audio, images, text or computer files, there are many forms of representation of this data, even a single type can have varying and very different type of representation.

For example an image in jpg format is composed of real numbers describing the luminance that is a grayscale representation of the image, and the chrominance which is the difference between the red color to the blue and green; then, there is the RGB representation that contains the intensity of the colors red, green and blue using only 8 bit integer numbers, another good example is text, that can easily be represented into and 8 bit ASCII variant. hence it is more efficient to work with these representations for the purpose of encryption.

3.1.1   Decimal and Binary Numbers

All types of data representations can be expressed as binary code, binary only contains two digits or bits 0 and 1, it is preferred in digital systems as it can be easily expressed with voltage with a certain level corresponding to 0 and another to 1, this simplifies its usage in processors and allows to make fairly complex operations even in low powered micro controllers.

The decimal system is represented by ten digits 0 to 9, each digit determines a quantity, the decimal system can represent non integer numbers by using a point to separate the integer part to the non integer, every number can be represented by a sum of base 10 numbers as follows:

Example 3.1 The integer number 65 can be written as a sum of base 10 numbers:

$$65 = \left(6 \times 10^2\right) + \left(5 \times 10^1\right)$$
$$= 60 + 5$$

Example 3.2 The number 32.01 is represented by the sum of base 10 numbers :

$$32.01 = \left(3 \times 10^1\right) + \left(2 \times 10^1\right) + \left(0 \times 10^{-1}\right) + \left(1 \times 10^{-2}\right)$$
$$= 30 + 2 + 0 + 0.01$$

Exercise 3.1 Represent the decimal number 256 as a sum of base 10 numbers.

Exercise 3.2 Represent the decimal number 8.456 as a sum of base 10 numbers.

Binary numbers are another way of representing magnitudes, and a much simpler than decimal since it only needs two digits or bits being them 0 and 1, having only two digits limits the quantity that can be represented, the maximum decimal number that can be represented with an binary is given by the next expression:

$$Decimal_{Max} = 2^n - 1$$

Where n is the number of binary bits.

Example 3.3 A 3 bit binary can quantify a decimal up to 7:

$$2^3 - 1 = 7$$

Example 3.4 A 8 bit binary can quantify a decimal up to 255:

$$2^8 - 1 = 255$$

Exercise 3.3 Determine the maximum decimal that can be represented with 16 binary bits.

Exercise 3.4 Determine the maximum decimal that can be represented with 32 binary bits.

The Table 3.1 gives a comparison of decimal and binary numbers:
Binary numbers are useful to comprehend operations that are very common in cryptography, for example the bitwise XOR logical operation denoted by $\oplus$:

Definition 3.1 The bitwise XOR operation between to binary digits is defined in Table 3.2

The property of bitwise XOR that makes it very useful in cryptography is shown in the next equations:

$$A \oplus B = C$$
$$A \oplus C = B$$
$$B \oplus C = A$$

Table 3.1 Decimal vs binary
numbers

Decimal	Binary
0	0
1	1
2	10
3	11
4	100
5	101
6	110
7	111
8	1000
9	1001
10	1010
11	1011
12	1100
13	1101
14	1110
15	1111

Table 3.2 XOR operation

A	B	$A \oplus B$
0	0	0
1	0	1
0	1	1
1	1	0

Table 3.3 Operation one's
complement

Digit	One's complement
0	1
1	0

Then $A \oplus B \oplus A = B$. The modulo operation is often used in encryption algorithms, to calculate it first it is necessary to introduce the one's complement and two's complement of a binary number.

Definition 3.2 The operation one's complement for a binary number is done by inverting the bits of the number as shown in Table 3.3

Example 3.5 Obtain the one's complement for the binary number 1010:

$$1010 \rightarrow 0101$$

Example 3.6 Obtain the one's complement for the binary number 1001 1001:

$$10011001 \rightarrow 01100110$$

Definition 3.3 The operation two's complement for a binary number is obtained by adding 1 to the less significant bit of the result of the one's complement operation.

Example 3.7 Obtain the two's complement for the binary number 1010: from previous examples the one's complements is 0101 then the twos complement is:

$$
\begin{array}{r}
0101 \\
+ \quad 1 \\
\hline
0110
\end{array}
$$

Example 3.8 Obtain the two's complement for the binary number 1001 1001: the one's complement for the expression is 01100110 so:

$$
\begin{array}{r}
01100110 \\
+ \quad 1 \\
\hline
01100111
\end{array}
$$

Example 3.9 Obtain the two's complement for 0110 and 01100111: the one's complement for each number is 1001 and 10011000 respectively, then:

$$
\begin{array}{r}
1001 \\
+ \quad 1 \\
\hline
1010
\end{array}
$$

And:

$$
\begin{array}{r}
10011000 \\
+ \quad 1 \\
\hline
10011001
\end{array}
$$

From the previous example it is possible to see that the operation has the same property as the bitwise XOR operation, denote two's complement as TC($\cdot$) then:

$$TC\,(A) = B$$

$$TC\,(B) = A$$

$$TC\,[TC\,(A)] = A$$

These two operations can be used to obtain the modulo 2^N operation, the modulo operation returns the remainder after a division, notice that for to integers A, B, N the modulo operation behaves as $A + B \mod 2^N = A + (j + 2^N) \mod 2^N$, this operation is also very popular in encryption algorithms. when used with negative values it has the same property of the bitwise XOR and two's complement.

Example 3.10 The result of the operation $-86 \mod 256$: the binary representation of 255 and -86 is 1111 1111 and $-0101\ 0110$ respectively, then adding:

$$\begin{array}{r} 1111\ 1111 \\ -\ 0101\ 0110 \\ \hline 1010\ 1001 \end{array}$$

Considering 1010 1001 the result of one's complement and applying two's complement:

$$\begin{array}{r} 1010\ 1001 \\ +\qquad\qquad 1 \\ \hline 1010\ 1010 \end{array}$$

Which is the result of the modulo operation 1010 1010 $= 170$ and $-86 \mod 256 = 170$

Example 3.11 The result of the operation $170 \mod -256$: the binary representation of -255 and 170 is $-1111\ 1111$ and 1010 1010 respectively, then by adding:

$$\begin{array}{r} -\ 1111\ 1111 \\ 1010\ 1010 \\ \hline -\qquad 0101\ 0101 \end{array}$$

Considering 0101 0101 the result of one's complement and applying two's complement:

$$\begin{array}{r} -0101\ 0101 \\ -\qquad\qquad 1 \\ \hline -\qquad 0101\ 0110 \end{array}$$

Which is the original value $-0101\ 0110 = -86$. This process is more complex than the bitwise XOR therefore it needs more computing power, though it is preferred by many authors.

3.1.2 Binary to Decimal and Decimal to Binary Conversions

As stated previously many forms of messages are made of real numbers integers or decimal, but in order to encrypt them it is necessary to make operations to its binary equivalent and then transform them back into the corresponding type of decimal to make the ciphertext. The conversion from binary to decimal is a very straightforward process. the conversion is done by multiplying each binary digit by its weight and then adding the result, the next examples show the process in a simple way:

Example 3.12 Calculate the decimal equivalent of the binary 1010:

$$Binary : 1\ 0\ 1\ 0$$

$$weight : 2^3\ 2^2\ 2^1\ 2^0$$

$$decimal = \left(1 \times 2^3\right) + \left(0 \times 2^2\right) + \left(1 \times 2^1\right) + \left(0 \times 2^0\right)$$

$$= 8 + 0 + 2 + 0$$

$$= 10$$

Example 3.13 The decimal equivalent of the binary 1010110:

$$Binary : 1\ 0\ 1\ 0\ 1\ 1\ 0$$

$$weight : 2^6\ 2^5\ 2^4\ 2^3\ 2^2\ 2^1\ 2^0$$

$$decimal = \left(1 \times 2^6\right) + \left(0 \times 2^5\right) + \left(1 \times 2^4\right) + \left(0 \times 2^3\right) + \left(1 \times 2^2\right) + \left(1 \times 2^1\right)$$

$$+ \left(0 \times 2^0\right)$$

$$= 64 + 0 + 16 + 0 + 4 + 2 + 0$$

$$= 86$$

Exercise 3.5 Obtain the decimal equivalent of the binary 1101 0110.

Exercise 3.6 Obtain the decimal equivalent of the binary 1101.0110.

Conversion from binary to integer is a little more challenging so two methods are given, the first is the sum of weights, where the decimal number is expressed as a sum of the largest possible multiples of two, the multiples are then transformed into its equivalent binary weight, the process is exemplified next:

Example 3.14 The binary equivalent of the decimal 11:

$$11 = 8 + 2 + 1$$
$$= 2^3 + 2^1 + 2^0$$
$$= \left(1 \times 2^3\right) + \left(0 \times 2^2\right) + \left(1 \times 2^1\right) + \left(1 \times 2^0\right)$$
$$Binary = 1011$$

Example 3.15 The binary equivalent of the decimal 23:

$$23 = 16 + 4 + 2 + 1$$
$$= 2^4 + 2^2 + 2^1 + 2^0$$
$$= \left(1 \times 2^4\right) + \left(0 \times 2^3\right) + \left(1 \times 2^2\right) + \left(1 \times 2^1\right) + \left(1 \times 2^0\right)$$
$$Binary = 10111$$

Exercise 3.7 Calculate the binary equivalent of the decimal 33.

Exercise 3.8 Obtain the binary equivalent of the decimal 86.

The other method is a succession of divisions by 2, the integer is divided by 2, the remainder is the bit and the result is divided in the next iteration, the process is repeated until the quotient of the division is 0, the most significant bit is the last resulting bit from the division and the least significant bit is the first, the next example illustrates the process

Example 3.16 Compute the binary equivalent of the decimal 11 with the division succession:

$$\frac{11}{2} = 5 \; remainder : 1$$
$$\frac{5}{2} = 2 \; remainder : 1$$
$$\frac{2}{2} = 1 \; remainder : 0$$
$$\frac{1}{2} = 0 \; remainder : 1$$
$$Binary \quad = \quad 1011$$

Example 3.17 Tthe binary equivalent of the decimal 23 with the division succession:

$$\frac{23}{2} = 11 \ remainder : 1$$

$$\frac{11}{2} = 5 \ remainder : 1$$

$$\frac{5}{2} = 2 \ remainder : 1$$

$$\frac{2}{2} = 1 \ remainder : 0$$

$$\frac{1}{2} = 0 \ remainder : 1$$

$$Binary \quad = \quad 10111$$

Exercise 3.9 Compute the binary equivalent of the decimal 33 with the division succession.

Exercise 3.10 Calculate the binary equivalent of the decimal 86 with the division succession.

This two methods are the most popular because they are simple and effective, the conversion of binary to decimal and decimal to binary is widely required when designing encryption algorithms.

3.1.3   Representation of Plaintext with 8 Integers

For ease of visualization in the book, in the following all messages are made of 8 bit integer data, the messages are to be restricted to text in 8 bit ASCII or images in RGB format, the 8 bit integer encompasses integer numbers with values ranging from 0 to 255 and allow to do operations with them with ease and low computing power consumption.

The text messages are structured as a vector of size $1 \times n$ consisting of 8 bit integers, each element of the vector is a character, an example of this is given next:

Example 3.18 Convert the plaintext "hello" into a 8 bit integer vector:

$$Plaintext : \quad h \quad e \quad l \quad l \quad o$$
$$vector \quad : \quad 104 \ 101 \ 108 \ 108 \ 111$$

the corresponding binary code is:

$$binary: \begin{bmatrix} 01101000 \\ 01100101 \\ 01101100 \\ 01101100 \\ 01101111 \end{bmatrix}^{T}$$

Example 3.19 Convert the plaintext "page9" into a 8 bit integer vector:

$$\begin{aligned} Plaintext & : & p & \quad a & \quad g & \quad e & \quad 9 \\ vector & : & 112 & \quad 97 & \quad 103 & \quad 101 & \quad 57 \end{aligned}$$

the corresponding binary code is:

$$binary: \begin{bmatrix} 01110000 \\ 01100001 \\ 01100111 \\ 01100101 \\ 00111001 \end{bmatrix}^{T}$$

Exercise 3.11 Convert the plaintext "sun" to a 8 bit integer vector and its binary equivalent.

Exercise 3.12 Convert the plaintext "Monday" to a 8 bit integer vector and its binary equivalent.

This representation makes easy to do logic operations with the plaintext and any other set of values of the same type:

Example 3.20 The result of the bitwise XOR operation between the plaintext "hello" and the phrase "12345":

$$\begin{aligned} Plaintext & : & h & \quad e & \quad l & \quad l & \quad o \\ Vector & : & 104 & \quad 101 & \quad 108 & \quad 108 & \quad 111 \\ Phrase & : & 1 & \quad 2 & \quad 3 & \quad 4 & \quad 5 \\ Vector & ; & 49 & \quad 50 & \quad 51 & \quad 52 & \quad 53 \end{aligned}$$

the operation is:

$$
\begin{bmatrix} 1101000 \\ 1100101 \\ 1101100 \\ 1101100 \\ 1101111 \end{bmatrix}^{T} \oplus \begin{bmatrix} 0110001 \\ 0110010 \\ 0110011 \\ 0110100 \\ 0110101 \end{bmatrix}^{T} = \begin{bmatrix} 1011001 \\ 1010111 \\ 1011111 \\ 1011000 \\ 1011010 \end{bmatrix}
$$

$$
\begin{bmatrix} 104 \\ 101 \\ 108 \\ 108 \\ 111 \end{bmatrix}^{T} \oplus \begin{bmatrix} 49 \\ 50 \\ 51 \\ 52 \\ 53 \end{bmatrix}^{T} = \begin{bmatrix} 89 \\ 87 \\ 95 \\ 88 \\ 90 \end{bmatrix}^{T}
$$

The ASCII representation of the solution is "YW_XZ".

Note that the plaintext was modified from its original form to be unreadable, this is a very basic form of encrypting or hiding a message, so the resulting ciphertext is "YW_XZ", in this example the encryption is done by a simple bitwise XOR operation using the key "12345" as the encryption value, then the cryptographic function is described by:

$$E\,(a,b) = c$$

$$E\,(a,b) = a \oplus b$$

$$E\,("hello", "12345") = "hello" \oplus "12345"$$

$$E\,("hello", "12345") = \text{``}YW\_XZ\text{''}$$

The encryption must have a way of decrypting the message or a decryption function $D(a,b)$, using the properties of the bitwise XOR allows the decryption function to be:

$$D\,(c,b) = c \oplus b$$

$$D\,(\text{``}YW\_XZ\text{''}, "12345") = \text{``}YW\_XZ\text{''} \oplus "12345"$$

$$D\,(\text{``}YW\_XZ\text{''}, "12345") = "hello$$

This is shown in the next example:

Example 3.21 Decrypt the ciphertext "YW_XZ" with the key "12345" by doing the bitwise XOR operation between them:

$$
\begin{aligned}
Ciphertext &: \quad Y \quad W \quad \_ \quad X \quad Z \\
Vector &: \quad 104 \; 101 \; 108 \; 108 \; 111 \\
key &: \quad 1 \quad 2 \quad 3 \quad 4 \quad 5 \\
Vector &; \quad 89 \quad 87 \quad 95 \quad 88 \quad 90
\end{aligned}
$$

The decryption function binary representation is:

$$
\begin{bmatrix} 1011001 \\ 1010111 \\ 1011111 \\ 1011000 \\ 1011010 \end{bmatrix}^{T}
\oplus
\begin{bmatrix} 0110001 \\ 0110010 \\ 0110011 \\ 0110100 \\ 0110101 \end{bmatrix}^{T}
=
\begin{bmatrix} 1101000 \\ 1100101 \\ 1101100 \\ 1101100 \\ 1101111 \end{bmatrix}^{T}
$$

The 8 bit integer operation is:

$$
\begin{bmatrix} 89 \\ 87 \\ 95 \\ 88 \\ 90 \end{bmatrix}^{T}
\oplus
\begin{bmatrix} 49 \\ 50 \\ 51 \\ 52 \\ 53 \end{bmatrix}^{T}
=
\begin{bmatrix} 104 \\ 101 \\ 108 \\ 108 \\ 111 \end{bmatrix}^{T}
$$

The message is correctly decrypted making the plaintext "hello" readable.

Exercise 3.13 Using the bitwise XOR operation, encrypt the plaintext "page9" with the key "67890".

Exercise 3.14 Using the bitwise XOR operation, decrypt the resulting ciphertext from the previous exercise.

3.1.4   Representation of Plain Images with 8 Bit Integers

Images can also be represented with 8 bit integers, digital images are a set of pixels arranged to form the image, each pixel contains information regarding intensity of color, luminance, chrominance or many other values depending on the image format, the contents and interpretation of the numerical value of the pixel also changes depending on the

Fig. 3.1 Grayscale image

format, but generally, images can be seen as a matrix of size $n \times m$ where each element of the matrix is a pixel and contains the numerical value of the said pixel corresponding to the format of the image. Consider the 8 bit integer 4×4 matrix:

$$Im = \begin{bmatrix} 0 & 0 & 64 & 64 \\ 0 & 0 & 64 & 64 \\ 128 & 128 & 255 & 255 \\ 128 & 128 & 255 & 255 \end{bmatrix}$$

To form a grayscale image with the matrix each one of its elements its considered the intensity of the white with respect to the black, then the value 0 is black and 255 is white, the resulting grayscale image has 32 pixels and its visual interpretation can be seen in Fig. 3.1.

Images can also be subjected to operations the same way that text messages, the difference is that this time the operation is done with another matrix containing the same type of data and having the same size, to exemplify this consider the 32 pixel grayscale image in the next example:

Example 3.22 Do a bitwise XOR operation between the grayscale image and the matrix K:

$$K = \begin{bmatrix} 50 & 50 & 100 & 100 \\ 50 & 50 & 100 & 100 \\ 150 & 150 & 200 & 200 \\ 150 & 150 & 200 & 200 \end{bmatrix}$$

Fig. 3.2 Grayscale operation
image

The operation is done element wise, then it is composed of 32 different bitwise XOR operations:

$$Ci = Im \oplus K$$

$$= \begin{bmatrix} 0 & 0 & 64 & 64 \\ 0 & 0 & 64 & 64 \\ 128 & 128 & 255 & 255 \\ 128 & 128 & 255 & 255 \end{bmatrix} \oplus \begin{bmatrix} 50 & 50 & 100 & 100 \\ 50 & 50 & 100 & 100 \\ 150 & 150 & 200 & 200 \\ 150 & 150 & 200 & 200 \end{bmatrix}$$

$$= \begin{bmatrix} 50 & 50 & 36 & 36 \\ 50 & 50 & 36 & 36 \\ 22 & 22 & 55 & 55 \\ 22 & 22 & 55 & 55 \end{bmatrix}$$

The resulting image is shown in Fig. 3.2.

The image has been considerably modified, this is because Images can be encrypted the same way text messages are with the bitwise XOR operation, the process of hiding images is very similar, the encryption function receives a plain image and the key, then it returns a cipher image. There is a decryption process related to the encryption, it is done the same way as in the text, the bitwise XOR operation is done element wise between the cipher image and the key, the encryption process is then:

$$E\,(Im, K) = Ci$$

$$E\,(Im, K) = Im_{n,m} \oplus K_{n,m}$$

Where $Im_{n.m}$ is the individual m,n pixel of the image and the $K_{n,m}$ is the corresponding element of the key, the decryption process is similar:

$$D\,(Ci,\,K) = Ci_{n,m} \oplus K_{n,m}$$
$$D\,(Ci,\,K) = Im_{n,m}$$

Once the operation is done the resulting image is the original plain image.

Example 3.23 Decrypt the cipher image of the previous example by doing the bitwise XOR operation between the cipher image Ci and the Key K:

$$Im = Ci \oplus K$$

$$= \begin{bmatrix} 50\ 50\ 36\ 36 \\ 50\ 50\ 36\ 36 \\ 22\ 22\ 55\ 55 \\ 22\ 22\ 55\ 55 \end{bmatrix} \oplus \begin{bmatrix} 50 & 50 & 100 & 100 \\ 50 & 50 & 100 & 100 \\ 150 & 150 & 200 & 200 \\ 150 & 150 & 200 & 200 \end{bmatrix}$$

$$= \begin{bmatrix} 0 & 0 & 64 & 64 \\ 0 & 0 & 64 & 64 \\ 128 & 128 & 255 & 255 \\ 128 & 128 & 255 & 255 \end{bmatrix}$$

The resulting image is shown in Fig. 3.3.

Fig. 3.3 Grayscale image after the decryption process is successful

The decryption is successful, note that if the encryption key is changed the decrypted image is not the plain image:

Example 3.24 Decrypt the cipher image of the previous example by doing the bitwise XOR operation between the cipher image Ci and the erroneous key K_2:

$$K = \begin{bmatrix} 200 & 200 & 150 & 150 \\ 200 & 200 & 150 & 150 \\ 100 & 100 & 50 & 50 \\ 100 & 100 & 50 & 50 \end{bmatrix}$$

The decryption with the wrong key is then:

$$Im = Ci \oplus K_2$$

$$= \begin{bmatrix} 50 & 50 & 36 & 36 \\ 50 & 50 & 36 & 36 \\ 22 & 22 & 55 & 55 \\ 22 & 22 & 55 & 55 \end{bmatrix} \oplus \begin{bmatrix} 200 & 200 & 150 & 150 \\ 200 & 200 & 150 & 150 \\ 100 & 100 & 50 & 50 \\ 100 & 100 & 50 & 50 \end{bmatrix}$$

$$= \begin{bmatrix} 250 & 250 & 178 & 178 \\ 250 & 250 & 178 & 178 \\ 114 & 114 & 5 & 5 \\ 114 & 114 & 5 & 5 \end{bmatrix}$$

The decryption is not successful due to the erroneous key and the message remains hidden as it is shown in Fig. 3.4

What has been presented is the basic idea of encryption, if the key is correct the ciphertext is correctly decrypted and the plaintext is the accessible to the user, if the key is not correct the ciphertext is not properly decrypted and the plaintext remains hidden.

Color images have many formats, one of the oldest and most well known is RGB, it is based on creating three images corresponding to the red, green and blue respectively, when the images are displayed together a color image is formed, the RGB format is based on the ability to create a wide range of colors using the three primary colors. A common form of representing images of size $n \times m$ pixels in the RGB format is to create tree matrices of size $n \times m$, each matrix contains the intensity of a primary color, then the image is composed of three matrices of size $n \times m$, where an element of the matrix contains the intensity of its corresponding primary color consisting in the range of 8 bit integer values from 0 to 255, the corresponding pixel of the color image is made of the combination of the 3 pixels of the red, green and blue matrices, an intensity of color 0 is displayed as black

Fig. 3.4 Grayscale image after
the decryption process fails

and a intensity of 255 is displayed as white, the next example shows how the structure of
an RGB image is done:

Example 3.25 Consider the next 8 bit integer matrices of size 2×2 corresponding to the
primary color components red R, green G and blue B:

$$R = \begin{bmatrix} 255 & 255 \\ 0 & 0 \end{bmatrix}, \; G = \begin{bmatrix} 255 & 0 \\ 255 & 0 \end{bmatrix}, \; B = \begin{bmatrix} 255 & 0 \\ 0 & 255 \end{bmatrix}$$

They are to combined into an RGB image of 2×2 pixels, the image is then presented
as an array of elements of size $2 \times 2 \times 3$ being $Im = [R, G, B]$, the image has four pixels:

$$Im = \begin{bmatrix} 1 & 2 \\ 3 & 4 \end{bmatrix}$$

Where the pixel one is white with color components R=255, G=255 and B=255. The
second pixel is red with components R=255, G=0 and B=0. The third pixel green with
R=0, G=255 and B=0. The fourth is blue with components R=0, G=0 and B=255. The
visual representation of this image is shown in Fig. 3.5.

The example may not show the full capacity of RGB, but it can be used to efficiently
represent large and colorful pictures of various resolutions.

This RGB format is very practical for the purpose of encryption, once again, being
it composed by 8 bit integer data makes very easy to make operations with the color
components, the bitwise XOR once again is one of the favorite functions for encryption,
the structure of the encryption is the same as in the case of the grayscale image, but this

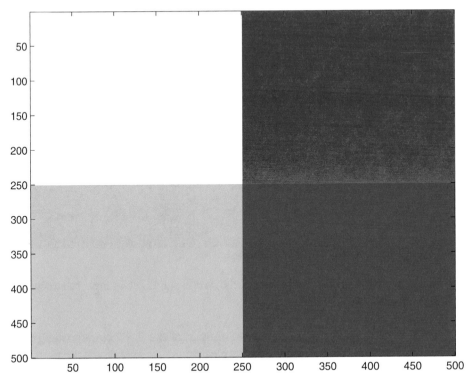

Fig. 3.5 Color image with RGB pixels: pixel 1 [255,255,255], pixel 2 [255,0,0], pixel 3 [0, 255,0] and pixel 4 [0 , 0 ,255]

time the encryption key must be larger to provide enough values to modify every pixel of the plain image.

Example 3.26 The following encryption Key=[20,40,60,80,100,120,140,160,180,200, 220,240] is to be used to encrypt the 4 pixel image of the previous example, the key is rearranged into three matrices denoted K_i:

$$K_1 = \begin{bmatrix} 20 \ 40 \\ 60 \ 80 \end{bmatrix}, \quad K_2 = \begin{bmatrix} 100 \ 120 \\ 140 \ 160 \end{bmatrix}, \quad K_3 = \begin{bmatrix} 180 \ 200 \\ 220 \ 240 \end{bmatrix}$$

The encryption is the same, the cryptographic function is:

$$E (Im, K) = C_i$$

$$E (Im, K) = Im \oplus K$$

Since the image has three color components, each section of the key denoted by K_i is used for one of the components:

$$E_R (R, K_1) = R \oplus K_1$$

$$E_G (G, K_1) = G \oplus K_2$$

$$E_B (B, K_1) = B \oplus K_3$$

The bitwise XOR operation is again done element wise, the numeric values are:

$$C_R = \begin{bmatrix} 255 & 255 \\ 0 & 0 \end{bmatrix} \oplus \begin{bmatrix} 20 & 40 \\ 60 & 80 \end{bmatrix},$$

$$C_R = \begin{bmatrix} 235 & 215 \\ 60 & 80 \end{bmatrix}$$

$$C_G = \begin{bmatrix} 255 & 0 \\ 255 & 0 \end{bmatrix} \oplus \begin{bmatrix} 100 & 120 \\ 140 & 160 \end{bmatrix}$$

$$C_G = \begin{bmatrix} 155 & 120 \\ 115 & 160 \end{bmatrix}$$

$$C_B = \begin{bmatrix} 255 & 0 \\ 0 & 255 \end{bmatrix} \oplus \begin{bmatrix} 180 & 200 \\ 220 & 240 \end{bmatrix}$$

$$C_B = \begin{bmatrix} 75 & 200 \\ 220 & 15 \end{bmatrix}$$

The cipher image is formed by the combination of the encrypted color components $C_{Im} = [C_R, C_G, C_B]$, the visual representation of the encrypted image is observed in Fig. 3.6.

The cipher image does not resemble the same colors as the plain image, although they bear the same size and structure but different values. It is possible to decrypt cipher image with the function:

$$D (C_{im}, K) = C_{im} \oplus K$$

$$D (C_{im}, K) = Im$$

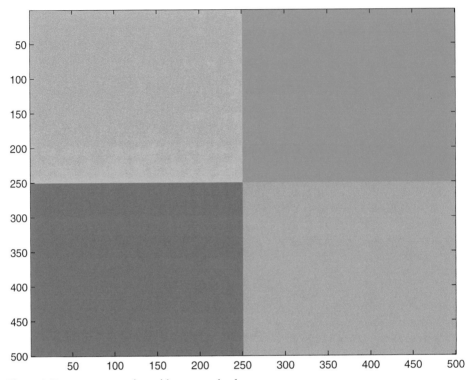

Fig. 3.6 Image representation with encrypted colors

Example 3.27 Using the decryption function with the key yields:

$$R = \begin{bmatrix} 235 & 215 \\ 60 & 80 \end{bmatrix} \oplus \begin{bmatrix} 20 & 40 \\ 60 & 80 \end{bmatrix}$$

$$R = \begin{bmatrix} 255 & 255 \\ 0 & 0 \end{bmatrix}$$

$$G = \begin{bmatrix} 155 & 120 \\ 115 & 160 \end{bmatrix} \oplus \begin{bmatrix} 100 & 120 \\ 140 & 160 \end{bmatrix}$$

$$G = \begin{bmatrix} 255 & 0 \\ 255 & 0 \end{bmatrix}$$

$$B = \begin{bmatrix} 75 & 200 \\ 220 & 15 \end{bmatrix} \oplus \begin{bmatrix} 180 & 200 \\ 220 & 240 \end{bmatrix}$$

$$B = \begin{bmatrix} 255 & 0 \\ 0 & 255 \end{bmatrix}$$

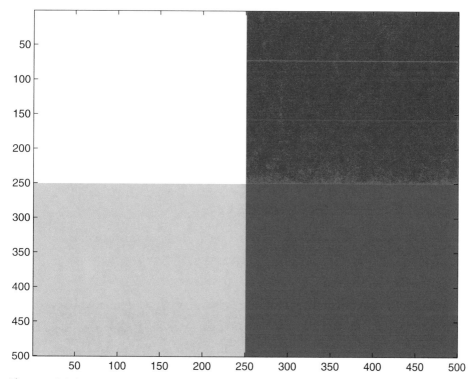

Fig. 3.7 Original color image after the decryption process

The decryption process is successful yielding the original plain image, shown in Fig. 3.7
If the image is decrypted with a wrong key the decryption process will create a different plain image than the one that was originally used.

Example 3.28 The cipher image is decrypted with the key $K = [1, 2, 3, 4, 5, 6, 7, 8, 9, 10, 11, 12]$, the key is arranged into the matrices:

$$K_1 = \begin{bmatrix} 1 & 2 \\ 3 & 4 \end{bmatrix}, \quad K_2 = \begin{bmatrix} 5 & 6 \\ 7 & 8 \end{bmatrix}, \quad K_3 = \begin{bmatrix} 9 & 10 \\ 11 & 12 \end{bmatrix}$$

And the decryption process yields the plain image:

$$R = C_R \oplus K_1$$

$$= \begin{bmatrix} 235 & 215 \\ 60 & 80 \end{bmatrix} \oplus \begin{bmatrix} 1 & 2 \\ 3 & 4 \end{bmatrix}$$

$$= \begin{bmatrix} 234 & 213 \\ 63 & 84 \end{bmatrix}$$

$$G = C_G \oplus K_2$$

$$= \begin{bmatrix} 155 & 120 \\ 115 & 160 \end{bmatrix} \oplus \begin{bmatrix} 5 & 6 \\ 7 & 8 \end{bmatrix}$$

$$= \begin{bmatrix} 158 & 126 \\ 116 & 168 \end{bmatrix}$$

$$B = C_B \oplus K_3$$

$$= \begin{bmatrix} 75 & 200 \\ 220 & 15 \end{bmatrix} \oplus \begin{bmatrix} 9 & 10 \\ 11 & 12 \end{bmatrix}$$

$$= \begin{bmatrix} 66 & 194 \\ 215 & 3 \end{bmatrix}$$

With the decrypted cipher image shown in Fig. 3.8

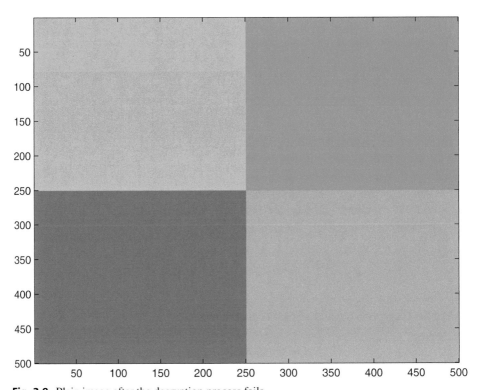

Fig. 3.8 Plain image after the decryption process fails

The decryption is unsuccessful and the plain image does not resemble the original message, this basic form of encryption shows how the modification of an image looks when encryption is done, also illustrates the results of failure to decrypt.

Exercise 3.15 Encrypt a 2×2 pixel color image with colors black $(0,0,0)$, yellow $(255,255,0)$, purple $(128,0,128)$ and orange $(255,165,0)$ using the key $K = [15, 25, 35, 45, 55, 65, 75, 85, 95, 105, 115, 125]$.

Exercise 3.16 Decrypt the cipher image from the previous exercise.

3.1.5 Data Carrier Signal

Many results in secure communications based on non linear control are stream ciphers, this encryption method usually needs to convert any message into a stream of values that are combined with another signal to hide the message, one of the most simple ways of making a signal that contains a message is to rearrange the ciphertext into a vector V_m, the values of the elements of the message vector indicate the amplitude that the signal takes at that given moment. Then create a time vector of the same size V_t, the time vector contains the moment when each element of the message vector is transmitted, every element of the time vector contains a numeric value in seconds. An alternative is to fix a time period and only specify the initial time for the message to be transmitted. The structure of the message and its corresponding time vector is:

$$V_m = S_0 \ S_1 \ S_2 \ \cdots \ S_n$$

$$V_t = t_0 \ t_1 \ t_2 \ \cdots \ t_n$$

Where S_0 is the first element of the message, S_n is the last element of the message, t_0 is the initial time and t_n is the final time of the transmission.

Example 3.29 The plaintext "hello" is to be converted into a data carrier signal, since the message is already in a vector form it is only necessary to associate a time vector:

$$plaintext : h\ e\ l\ l\ o$$

$$V_m = \begin{bmatrix} 104 \ 101 \ 108 \ 108 \ 111 \end{bmatrix}$$

$$V_t = \begin{bmatrix} 1 \ 2 \ 3 \ 4 \ 5 \end{bmatrix}$$

The initial time of the vector is $t_0 = 1$ and the transmission of the message finishes at $t_4 = 5$, the resulting signal is shown in Fig. 3.9

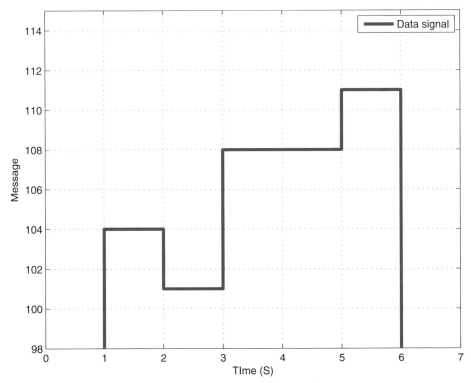

Fig. 3.9 Ciphertext data signal

The received signal is then converted into a vector recovering the original message "hello".

Images can be turned into data carrier signals the same way, but with them it is necessary to transform the R,G and B matrices into the vector V_m, also the transmission time is considerably larger than the text messages.

Example 3.30 Using the image from Fig. 3.5 the vector is arranged as follows:

$$V_t = \begin{bmatrix} V_R & V_G & V_B \end{bmatrix}$$

Where

$$V_R = \begin{bmatrix} 255 & 255 & 0 & 0 \end{bmatrix}$$

$$V_G = \begin{bmatrix} 255 & 0 & 255 & 0 \end{bmatrix}$$

$$V_B = \begin{bmatrix} 255 & 0 & 0 & 255 \end{bmatrix}$$

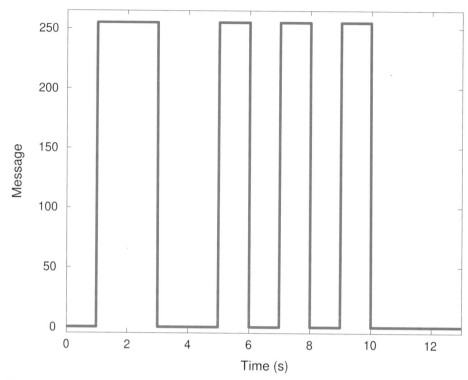

Fig. 3.10 Cipher image signal

The time vector is:

$$v_t = \begin{bmatrix} 1\ 2\ 3 \cdots 12 \end{bmatrix}$$

The message is transmitted the same way, with the data carrier signal commencing at time $t_0 = 1$ seconds and ending at $t_{11} = 12$ seconds (see Fig. 3.10).

The data carrier signal is considerably larger but also, the message contains more information.

Exercise 3.17 Convert the ciphertext "YW_XZ" into a signal and transmit it with initial time $t_0 = 1$ with a time period of 0.1 s.

Exercise 3.18 Convert the cipher image of Figs. 3.4, 3.5, and 3.6 into a signal and transmit it with initial time $t_0 = 0.2$ and period of 0.4 s.

There are other variations of message conversion to signal, another very popular variant is binary signals that are made of binary values, this method is preferred in many stream cyphers for its simplicity and resilience to noise, but it has the setback that either the needed transmission time is multiplied by eight or the signal period is decreased by a factor of 8, requiring faster data acquisition, this method will be discussed in detail in following chapters.

3.2   Stream Ciphers and State Observers

Stream ciphers are the most common cipher found in the literature about secure communications based on non-linear control as their implementation with synchronization is straight forward, since operate with relative low computer complexity and at high speeds they are ideal for the encryption of large messages, they work by encrypting plaintext one element at a time, the element size can range from a bit to integers, decimals or any other type of data, depending on the composition of the plaintext. The most simple form of stream cipher is called keystream generator or running key generator, the idea is to combine a plaintext signal with the values of the keystream formed by pseudorandom numbers, this keystream is formed by an algorithm, usually dependent on the key. The combination of the plaintext with the keystream is usually done with a bitwise XOR operation. Let the plaintext bits be $p_1, p_2, \ldots, p_i$ and the keystream bits $ks_1, ks_2, \ldots, ks_i$, the ciphertext for the plaintext is

$$c_i = p_i \oplus ks_i$$

The decryption process is similar, the bitwise XOR properties make possible that $a \oplus b = c$ then $c \oplus a = b$ and $c \oplus b = a$, by making use of this property, the decryption needs only to apply bitwise XOR to the known keystream and the ciphertext to recover the plaintext:

$$p_i = ks_i \oplus c_i$$

The image in Fig. 3.11 shows the process.

Other type of very popular stream cipher are the self-synchronizing stream ciphers or ciphertext auto key (CTAK), the idea of this stream cipher is to increase the security

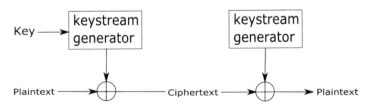

Fig. 3.11 Representation of the stream cipher process

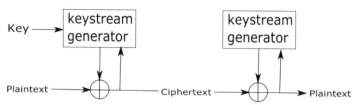

Fig. 3.12 Representation of the ciphertext auto key (CTAK) process

of the common stream cipher by making the current keystream depend on the previous ciphertext, the encryption and decryption processes are identical to the previous stream cipher, the only difference lies in the generation of the keystream, that now depends on the ciphertext and the key (see Fig. 3.12).

A problem of the CTAK is the possibility of error propagation, if an erroneous ciphertext bit is used in the decryption process it generates an erroneous decryption and, depending on how the keystream generator works and how is implemented, the error could propagate to other ciphertext bits decryption, damaging the decrypted data. In both cases a very important component of the stream cipher is the pseudorandom numbers used as keystream, given that the random generator is reliable, the algorithm can confidently use it, but if it has problems, these will carry on the encryption algorithm and make it unreliable.

3.2.1   Pseudorandom Number Generator

An important component for any cipher is a pseudorandom number generator (PRNG), this generator creates vectors of pseudorandom numbers that are used as input for the permutation and mixing operations, a simple to implement PRNG is called Blum Blum Shub [3], it was introduced in 1986 by Lenore Blum, Manuel Blum and Michael Shub, it is easy to implement and has low computational complexity, it is described by:

$$r_{n+1} = r_n^2 \ \text{mod} \ pq$$

Where r_n is a pseudorandom number and p and q are large prime numbers,

Example 3.31 Using $r_0 = 54$, $p = 11$, $q = 11$ and $n = 10$ create 10 pseudorandom numbers with the Blum Blum Shub PRNG (see Table 3.4).

Exercise 3.19 Using $r_0 = 5$, $p = 11$, $q = 19$ and $n = 10$ create 10 pseudorandom numbers with the Blum Blum Shub PRNG.

Exercise 3.20 Using $r_0 = 123456$, $p = 11$, $q = 19$ and $n = 10$ create 10 pseudorandom numbers with the Blum Blum Shub PRNG.

Table 3.4 Pseudorandom
numbers creation with the
Blum Blum Shub PRNG

n	r_{n+1}	r_n
1	$54^2 \bmod 121$	12
2	$12^2 \bmod 121$	23
3	$23^2 \bmod 121$	45
4	$45^2 \bmod 121$	89
5	$89^2 \bmod 121$	56
6	$56^2 \bmod 121$	111
7	$111^2 \bmod 121$	100
8	$100^2 \bmod 121$	78
9	$78^2 \bmod 121$	34
10	$34^2 \bmod 121$	67

The synchronization of two chaotic systems can be used for generating random numbers by including the chaotic states into the dynamic of the Blum Blum Shub PRNG. The process is as follows:

1. Form the initial conditions for the chaotic oscillator from the key, every state needs a different key element which is normalized between 0 and 1, a value that is easily scaled into the needed range that falls within the chaotic attractor:

$$x_1(0) = x_1(0) x_1(0) \frac{k_1}{99999}, x_2(0) = \frac{k_2}{99999}, \ldots, x_n(0) = \frac{k_n}{99999}$$

2. Generate a long trajectory then obtain as many samples as pseudorandom numbers are needed, starting at time t_{0k} and period ω_k that depend on an element of the key:

$$t_{0k} = 5 + \frac{k_3}{99999}$$

$$\omega_k = \frac{k_4}{99999}$$

Check the parity of another element of the key $k_3 + k_4$ to determine if the period ω_k is in milliseconds or microseconds, the starting t_{0k} time is recommended to remain in seconds. The samples are then stored into a vector $V_c = \begin{bmatrix} V_{c1} & V_{c2} & \cdots & V_{cn} \end{bmatrix}$.

3. Normalize the vector V_C into values between 0 and 1, the new normalized value of V_{ci} is computed by:

$$V_{ci} = \frac{V_{ci} - \min(V_c)}{\max(V_c) - \min(V_c)}$$

Then the vector is transformed into a 16 bit integer vector:

$$V_{c16} = \text{round}\,(65535\,V_c)$$

4. Apply prime factorization to an element V_{c16i}, select the two largest prime numbers and assign them to p and q, if the element is 0 or 1 use 11 and 19 instead respectively, a key element k_5 determines de number of iterations for the Blum Blum Shub PRNG, finally the value V_{c16i} is used as the starting value r_0 for the PRNG and store the resulting number into the vector V_r.
5. Normalize the vector V_r between 0 and 1:

$$V_{ri} = \frac{V_{ri} - \min\,(V_r)}{\max\,(V_r) - \min\,(V_r)}$$

Example 3.32 A set of 10 pseudorandom numbers is generated with the key $K_{ey} = 12345 - 67891 - 01112 - 13141 - 51617$ and the Duffing oscillator:

$$\dot{x}_1 = x_2$$
$$\dot{x}_2 = -ax_2 - bx_1 - cx_1^3 + d\cos\omega ts$$
$$y = x_1$$

With the values $a = 0.2, b = -1, c = 1, d = 0.3, \omega = 1, g = 1/255$.

1. The initial conditions are:

$$x_1\,(0) = \frac{12345}{99999}$$
$$= 0.123451$$
$$x_2\,(0) = \frac{67891}{99999}$$
$$= 678916$$

And make the trajectories shown in Fig. 3.13.
2. The value $k_3 + k_4 = 01112 + 12131 = 14253$ is odd, then initial time and the period for the samples are:

$$t_{0k} = 5 + \frac{01112}{99999}$$
$$= 5.01112\,\text{s}$$
$$\omega_k = \frac{13141}{99999}$$
$$= 0.13141\,\mu\text{s}$$

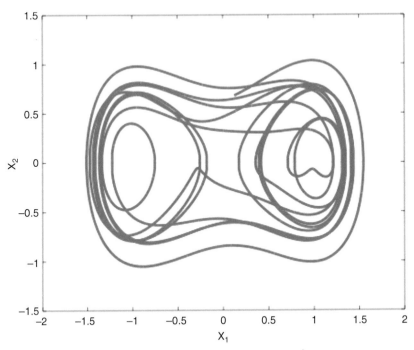

Fig. 3.13 State trajectories of the attractor for the Duffing oscillator

3. The sampled values are:

$$
V_c = \begin{bmatrix} 0.4672 \\ -1.3929 \\ 1.0474 \\ 1.3020 \\ 0.5362 \\ 0.7732 \\ 0.7294 \\ -0.3233 \\ 0.4664 \\ -0.9864 \end{bmatrix}^{T} \rightarrow V_c = \begin{bmatrix} 0.6557 \\ 0.0357 \\ 0.8491 \\ 0.9340 \\ 0.6787 \\ 0.7577 \\ 0.7431 \\ 0.3922 \\ 0.6555 \\ 0.1712 \end{bmatrix}^{T} \rightarrow V_{c16} = \begin{bmatrix} 42974 \\ 2340 \\ 55648 \\ 61209 \\ 44481 \\ 49658 \\ 48701 \\ 25705 \\ 42957 \\ 11219 \end{bmatrix}^{T}
$$

4. -The prime factors of k_5 are $71 \times 727 = 51617$ the values p=71 and q=727 along V_{c16} are the input for the BBS PRNG which is iterated 51617 times making:

$$V_{rn+1} = v_{r0}^2 \mod pq$$

$$V_{r0} = V_{c16i}$$

$$V_r = \begin{bmatrix} 28753 \\ 25005 \\ 50168 \\ 52113 \\ 12247 \\ 32097 \\ 29201 \\ 42356 \\ 46488 \\ 49458 \end{bmatrix}^T$$

Exercise 3.21   Obtain 10 pseudorandom numbers with the Van Der Pol oscillator

$$\dot{x}_1 = x_2$$
$$\dot{x}_2 = \mu \left(1 - x_1^2\right) x_2 - x_1$$
$$y = x_1$$

Exercise 3.22   Obtain 10 pseudorandom numbers with the Chua oscillator:

$$\dot{x} = -y - z$$
$$\dot{y} = x + ay$$
$$\dot{z} = b + z(x - c)$$

3.2.2   The Luenberger Observer in a Stream Cipher

When implementing stream ciphers with state observers the chaotic system to be observed takes the place of the key stream generator and the key stream is either the output or one of the states, the plaintext is then scaled down to be better hidden and combined with the key stream state by a sum. The decryption is done by the observer, it reconstructs the states and with them the output or whatever trajectory was used as key stream, finally the decryption is done by subtracting the estimated or reconstructed key stream obtaining the plaintext, the first encryption with a Luenberger observer [4, 5] is:

$$\dot{x} = Ax + f(x) + kgs$$

$$y = Cx + gs$$

$$\dot{\hat{x}} = A\hat{x} + f(\hat{x}) + k(y - \hat{y})$$

$$\hat{y} = C\hat{x}$$

$$\hat{s} = \frac{1}{g}(y - \hat{y})$$

Where s is the plaintext, $0 < g < 1$ is a scaling factor chosen to make the signal as small as possible and Cx is key stream. It could be confusing to add the gain vector and the message to the states of the chaotic system, but it is necessary for the proof of stability, the Lyapunov function is chosen as:

$$V = e^T P e$$

The function's derivative is:

$$\dot{V} = \dot{e}^T P e + e^T P \dot{e}$$

The derivative of the error function is expressed in terms of the states:

$$\dot{e} = \dot{x} - \dot{\hat{x}}$$

$$= Ax + f(x) + kgs - A\hat{x} - f(\hat{x}) - k(y - \hat{y})$$

$$= A(x - \hat{x}) + f(x) - f(\hat{x}) + kgs - k(y - \hat{y})$$

$$= Ae + f(x) - f(\hat{x}) + kgs - k(Cx + gs - C\hat{x})$$

$$= Ae + f(x) - f(\hat{x}) - kC(x - \hat{x}) + kgs - kgs$$

$$= Ae + f(x) - f(\hat{x}) - kCe$$

When defining $\phi(e) = f(x) - f(\hat{x})$ with the same assumptions as in the proof of stability of the Chap. 2 Luenberger observer, the error derivative is:

$$\dot{e} = Ae + \phi(e) - kCe$$

The derivative of the Lyapunov function is:

$$\dot{V} = [Ae + \phi\,(e) - kCe]^T\,Pe + e^T P\,[Ae + \phi\,(e) - kCe]$$
$$= e^T A^T Pe + \phi\,(e)^T\,Pe - (kCe)^T\,Pe + e^T PAe + e^T P\phi\,(e) - e^T PkCe$$
$$= e^T A^T Pe + \phi\,(e)^T\,Pe + e^T PAe + e^T P\phi\,(e) - (kCe)^T\,Pe - e^T PkCe$$

And if the assumptions are fulfilled

$$\dot{V} \le e^T A^T Pe + e^T PAe + 2\alpha e^T Pe + e^T e - (KCe)^T\,Pe - e^T PKCe$$
$$\le e^T \left[A^T P + PA + 2\alpha P + I\right] e - 2e^T PKCe$$

It yields to:

$$\dot{V} < e^T\,[Q - PKC]\,e$$

Showing that the states of the observer converge to the states of the chaotic system, in consequence the error of the message also converges to zero:

$$e_s = s - \hat{s}$$
$$= s - \frac{1}{g}\,(y - \hat{y})$$
$$= s - \frac{1}{g}\,(Cx + gs - C\hat{x})$$

The asymptotic stability of the error makes possible, after some time, to assume that $x = \hat{x}$ making:

$$e_s = s - \frac{1}{g}\,(gs)$$
$$= s - s$$
$$e_s = 0$$

Proving that the error also converges to zero as the states of the observer converge to the states of the chaotic system, hence the message is recovered.

Example 3.33 Encrypt an image by using the output of the Van der Pol oscillator as key stream and decrypt with the Luenberger observer. The Van der Pol oscillator with the embedded message is:

$$\dot{x}_1 = x_2 + k_1 g s$$

$$\dot{x}_2 = \mu \left(1 - x_1^2\right) x_2 - x_1 + k_2 g s$$

$$y = x_1 + g s$$

A Luenberger observer for the system has the following state equations:

$$\dot{\hat{x}}_1 = \hat{x}_2 + k_1 \left(y - \hat{y}\right)$$

$$\dot{\hat{x}}_2 = \mu \left(1 - \hat{x}_1^2\right) \hat{x}_2 - \hat{x}_1 + k_2 \left(y - \hat{y}\right)$$

$$y = \hat{x}_1$$

$$\hat{s} = y - \hat{y}$$

Using the values $\mu = 5$, $g = 1/255$, $k_1 = 3$ and $k_2 = 6$ the synchronization goes as follows. The message is shown in Fig. 3.14. The states convergence is observed in

Fig. 3.14 Original message to be encrypted by the Van der Pol oscillator

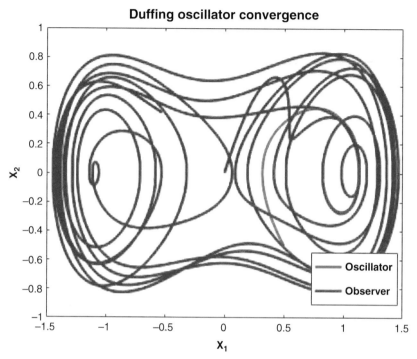

Fig. 3.15 Van der Pol oscillator and Luenberger observer convergence for decryption of the message

Fig. 3.15. The message recovery error is then shown in Fig. 3.16 and the encrypted image is shown in Fig. 3.17. The decrypted image is shown in Fig. 3.18.

Example 3.34 Encrypt an image by using the output of Colpitts oscillator as key stream and decrypt with the Luenberger observer. The message is shown in Fig. 3.19. The states convergence is in Fig. 3.20. The message recovery error is shown in Fig. 3.21 and the encrypted image is in Fig. 3.22. With the decrypted message shown in Fig. 3.23.

Example 3.35 Use the Rössler chaotic oscillator to encrypt another message. The message is shown in Fig. 3.24. The states convergence is shown in Fig. 3.25. The message recovery error is then shown in Fig. 3.26 and the encrypted image is shown in Fig. 3.27. With the decrypted message observed in Fig. 3.28.

Exercise 3.23 Use the Chua oscillator to encrypt the plain image formed by the color component matrices:

$$R = \begin{bmatrix} 255 & 255 \\ 0 & 0 \end{bmatrix}, \; G = \begin{bmatrix} 255 & 0 \\ 255 & 0 \end{bmatrix}, \; B = \begin{bmatrix} 255 & 0 \\ 0 & 255 \end{bmatrix}$$

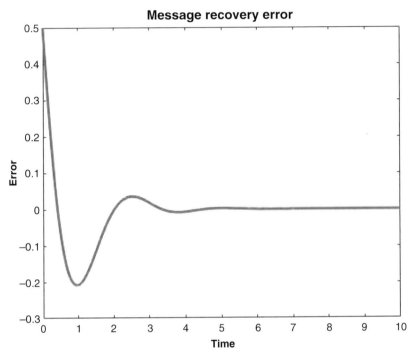

Fig. 3.16 Message recovery error by using the Van der Pol oscillator and the Luenberger observer

The second method for stream cipher encryption is to use as a key stream a state different than the ones that form the output, the encryption is also done by adding the plaintext to the key stream, the next equation represent the method:

$$\dot{x} = Ax + f(x)$$

$$y = Cx$$

$$y_s = C_2 x + gs$$

$$\dot{\hat{x}} = A\hat{x} + f(\hat{x}) + k(y - \hat{y})$$

$$\hat{y} = C\hat{x}$$

$$\hat{s} = \frac{1}{g}(y_s - C_2\hat{x})$$

Where y_s is ciphertext, the vector C_2 is used to select the state to use as key stream and it must be different that C. The output y is used to synchronize the states, note that this scheme does not change the proof of stability since the creation and transmission of the ciphertext is not a disturbance nor an uncertainty.

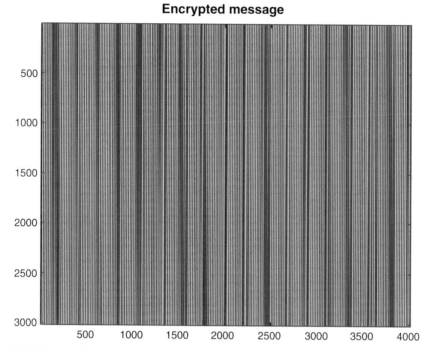

Fig. 3.17 Encrypted message by the output of the Van der Pol oscillator

3.3   Block Ciphers and Observers

Block ciphers are the other major type of cipher that employ symmetric keys, these ciphers operate by applying an invariant cryptographic function to a set of elements of a plaintext of fixed size, these sets are named blocks and their size range from 32 bits to 64 bits usually, hence the name block cipher.

There are many successful examples of these type of ciphers such as AES, DES or RC5 which are quite intricate, for simplicity the block cipher presented in this chapter is made by some of the operations that compose those algorithms, in particular a permutation and mixing operation will be done, the remaining operations depend, mostly, on search boxes thus will not be included.

This time the Synchronization will not be directly involved into the encryption, it will rather be used as a source for a pseudorandom number generator that creates data required as input for the permutation and mixing operations.

The most important component of the algorithm is the key, as the security must depend exclusively on it, also it is used in many of the operations of the algorithm, this chapter will utilize user defined keys of fixed size, the most easy method is a numeric key divided

Recovered message

Fig. 3.18 Recovered message by the Luenberger observer after the encryption of the Van der Pol oscillator

into small key sections denoted k_n, $n \in \mathbb{N}$:

$$K_{ey} = k_1 - k_2 - k_3 - \ldots - k_n$$

An example key is:

$$K_{ey} = 12345 - 12345 - 12345 - \ldots - 12345$$

Keys usually contain ASCII values, but for simplicity of representation the keys in this chapter are restrained to decimal numbers with key sections consisting of 5 digit numbers.

Stream ciphers convert plaintext into ciphertex, this operation is done at the rate of 1 bit at a time.

Message

Fig. 3.19 Original message to be encrypted by the Colpitts oscillator

3.3.1   Block Cipher

A master system generates a chaotic trajectory based on initial conditions that depend on a key, the trajectories are used to generate a sequence of pseudorandom numbers that are the input for a block cipher, for decryption a slave system reconstructs said trajectories which are then used to reconstruct the random numbers, the set of reconstructed numbers are used to do the needed operations for the decryption.

The encryption process on the plaintext block is composed of two operations:

1. Mixing: the plaintext block is combined with a set of pseudorandom numbers of the same size of the block, the operation is made by a simple bitwise XOR operation:

$$CB_i = PB_i \oplus r_{ni}$$

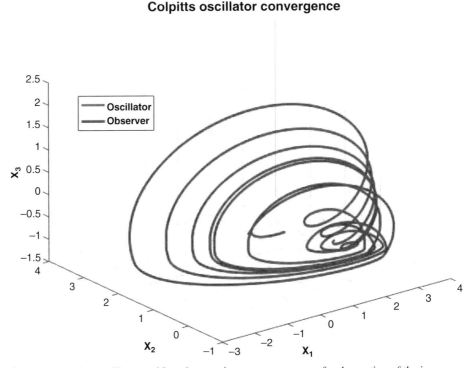

Fig. 3.20 Colpitts oscillator and Luenberger observer convergence for decryption of the image

Where CB is a partially encrypted plaintext block, PB is the plaintext block and r_n is a pseudorandom number.

2. Permutation: A set of pseudorandom numbers of the size of a plaintext block is ordered from smallest to largest, then the resulting message from the mixing step is reordered in the same way, the process is exemplified:

$$\begin{matrix} CB \\ Order \end{matrix} = \begin{bmatrix} a\ b\ c\ d\ e\ f\ g\ h \\ 1\ 2\ 3\ 4\ 5\ 6\ 7\ 8 \end{bmatrix}$$

$$\begin{matrix} PRN \\ Order \end{matrix} = \begin{bmatrix} 40\ 248\ 244\ 24\ 104\ 36\ 108\ 202 \\ 1\quad 2\quad 3\quad 4\quad 5\quad 6\quad 7\quad 8 \end{bmatrix}$$

$$\begin{matrix} PRN \\ New Order \end{matrix} = \begin{bmatrix} 24\ 36\ 40\ 104\ 108\ 202\ 244\ 248 \\ 4\quad 6\quad 1\quad 5\quad 7\quad 8\quad 3\quad 2 \end{bmatrix}$$

$$CTB = \begin{bmatrix} d\ f\ a\ e\ g\ h\ c\ b \\ 4\ 6\ 1\ 5\ 7\ 8\ 3\ 2 \end{bmatrix}$$

The ciphertext block is then CTB.

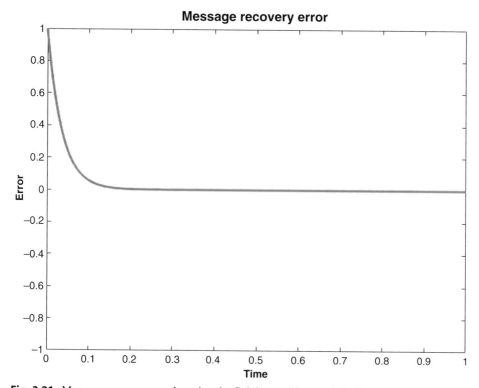

Fig. 3.21 Message recovery error by using the Colpitts oscillator and the Luenberger observer

The decryption process needs for a state observer to synchronize to the oscillator and with the key reconstruct the random numbers, then the encryption process is done backwards:

1. Reorder the ciphertext block CTB according to the reconstructed pseudorandom numbers $P\hat{R}N$:

$$CTB = \begin{bmatrix} d & f & a & e & g & h & c & b \\ 4 & 6 & 1 & 5 & 7 & 8 & 3 & 2 \end{bmatrix}$$

$$P\hat{R}N = \begin{bmatrix} 40 & 248 & 244 & 24 & 104 & 36 & 108 & 202 \\ 1 & 2 & 3 & 4 & 5 & 6 & 7 & 8 \end{bmatrix}$$

$$P\hat{R}N = \begin{bmatrix} 24 & 36 & 40 & 104 & 108 & 202 & 244 & 248 \\ 4 & 6 & 1 & 5 & 7 & 8 & 3 & 2 \end{bmatrix}$$

$$\hat{C}B = = \begin{bmatrix} a & b & c & d & e & f & g & h \\ 1 & 2 & 3 & 4 & 5 & 6 & 7 & 8 \end{bmatrix}$$

Encrypted message

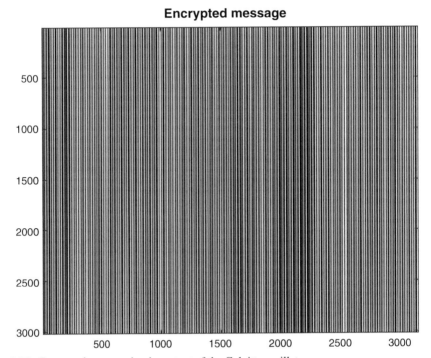

Fig. 3.22 Encrypted message by the output of the Colpitts oscillator

2. With the key reconstruct the values r_n and do the bitwise XOR operation to the reconstructed partially encrypted ciphertext block $\hat{C}B$:

$$PB = \hat{C}B \oplus r_n$$

Finally recovering the plaintext block:

Example 3.36 The plaintext "blockcipher " is encrypted with the key $K_{ey} = 12345 - 67891 - 01112 - 13141 - 51617$ and the duffing oscillator:

$$\dot{x}_1 = x_2$$
$$\dot{x}_2 = -ax_2 - bx_1 - cx_1^3 + d\cos\omega ts$$
$$y = x_1$$

Fig. 3.23 Recovered message by the Luenberger observer after the encryption of the Colpitts oscillator

With the values $a = 0.2, b = -1, c = 1, d = 0.3, \omega = 1, g = 1/255$. the pseudorandom numbers are arranged into to vectors, one for each step:

$$r_{n1} = \begin{bmatrix} 110 \\ 232 \\ 46 \\ 67 \\ 37 \\ 35 \\ 222 \\ 148 \\ 140 \\ 37 \\ 218 \end{bmatrix}, r_{n2} = \begin{bmatrix} 230 \\ 241 \\ 125 \\ 125 \\ 86 \\ 230 \\ 94 \\ 28 \\ 199 \\ 99 \\ 62 \end{bmatrix}$$

The first step produces the partially encrypted message that is then reordered by the second vector producing the ciphertext C".!2sL/0ç@.

Message

Fig. 3.24 Original message to be encrypted by the Rössler oscillator

The previous message is decrypted with the Luenberger observer.

$$\dot{\hat{x}}_1 = \hat{x}_2 + k_1 Ce$$

$$\dot{\hat{x}}_2 = -a\hat{x}_2 - b\hat{x}_1 - c\hat{x}_1^3 + d \cos \omega ts + k_2 Ce$$

$$\hat{y} = \hat{x}_1$$

The reconstructed random numbers are:

$$\hat{r}_{n1} = \begin{bmatrix} 11 \\ 43 \\ 166 \\ 187 \\ 165 \\ 115 \\ 139 \\ 76 \\ 190 \\ 48 \\ 175 \end{bmatrix}, \ \hat{r}_{n2} = \begin{bmatrix} 103 \\ 25 \\ 34 \\ 240 \\ 244 \\ 147 \\ 15 \\ 60 \\ 90 \\ 209 \\ 4 \end{bmatrix}$$

Rössler Oscillator convergence

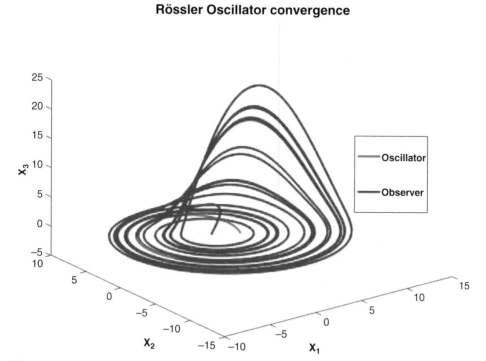

Fig. 3.25 Rössler oscillator and Luenberger observer convergence for decryption on the image

The reordered ciphertext is: "blockcipher ".

Exercise 3.24 Encrypt and decrypt the message "hello" using the Van der Pol oscillator.

Exercise 3.25 Encrypt and decrypt the message "random" using the Rössler oscillator.

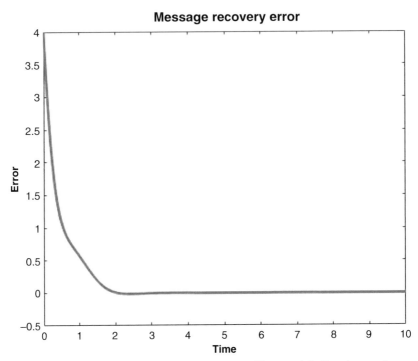

Fig. 3.26 Message recovery error by using the Rössler oscillator and the Luenberger observer

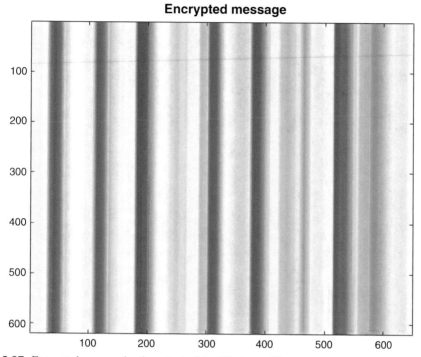

Fig. 3.27 Encrypted message by the output of the Rössler oscillator

Recovered Message

Fig. 3.28 Recovered message by the Luenberger observer after the encryption of the Rössler oscillator

References

1. Schneier, B. (2007). *Applied cryptography: Protocols, algorithms, and source code*. Hoboken: Wiley.
2. Rueppel, R. A. (2012). *Analysis and design of stream ciphers*. Berlin: Springer.
3. Blum, L., Blum, M., & Shub, M. (1986). A simple unpredictable pseudorandom number generator. *SIAM Journal on Computing, 15*(2), 364–383.
4. Chen, C. T. (1999). *Linear system theory and design*. Oxford: Oxford University Press.
5. Khalil, H. (1996). *Nonlinear systems*. Hoboken: Prentice Hall.

Liouvillian Systems and Cryptography

4

Abstract

In this chapter we show the use of a Liouvillian chaotic system on a master-slave scheme for secure communications. Two forms of data encryption are presented, the first is a state observer based on the Super-Twisting algorithm, the second is a receiver designed through the properties of Liouvillian systems, this last receiver does not suffer from data loss due to synchronization error present on state observer based encryption. To test the advantages of this encryption, images and text will be used as transmitted data.

4.1 Introduction

The problem of synchronization in chaotic systems has received a great attention among scientist in many fields due to its potential applications [1, 2], one of the most important applications is secure communications where a predominant technique is chaotic masking [3–6]; but also chaotic system alone have been used in secure communications [7, 8] and as random number generator due to its properties [9], also chaotic systems are present in other areas in the literature [10, 11]. In the last decade, many different approaches related to chaos synchronization have been applied, where the goal is to design an observer to achieve synchronization to nonlinear electric circuit oscillators [12]. For the synchronization problem a chaotic system is considered, under master-slave paradigm, where the transmitter is the master system and the receiver is the slave system, the objective is to synchronize the complete response of the slave system to the master system by driving the slave with a signal provided by the master [13].

© The Author(s), under exclusive license to Springer Nature Switzerland AG 2023
R. Martínez-Guerra et al., *Encryption and Decryption Algorithms for Plain Text and Images using Fractional Calculus*, Synthesis Lectures on Engineering, Science, and Technology, https://doi.org/10.1007/978-3-031-20698-6_4

Using state observers as receivers and message recovery systems has certain problems, the most important one is that they do not allow the creation of encryption keys from initial conditions, parameters or gains [14], as they are designed to converge to the transmitter states regardless of small variations in its values, this makes that the security of the data depends entirely on the masking provided by the transmitter which, in most cases is not enough [13]. The other important problem is that they present error in the recovered data, the magnitude of the error will depend on the observer and its design, but still, it will be present. A solution for these two problems is given by designing the receiver not as an observer but as a reconstruction of the states of the transmitter involved in data transmission, there is a class of systems that allow to do this kind of reconstruction without the need of state observers nor a full reconstruction of transmitters dynamical equations, they will be presented in the following sections along with the solution of this problems.

Two approaches for data encryption will be presented, the first is a state observer based on the Super-Twisting algorithm [15], the observer can reconstruct all the states of the transmitter and also recover the encrypted data, this receiver has both the problems mentioned above, but it has the advantage that the end user will not require full knowledge of the transmitters dynamic equations. The second solution is a reconstruction of the states that are involved in data transmission, this receiver will retain the security features of the Super-Twisting observer and it will also provide a solution to its lack of security and accuracy in data reconstruction. The two encryption methods will be tested with image and text as data, this will allow to present the advantages of the use of Liouvillian systems in secure communications.

This chapter is organized as follows: in Sect. 4.2 we present the transmitter and the class of systems that will be used for creating the receivers, in Sect. 4.3 the receivers and their unique characteristics are given, in Sect. 4.4 we give the numerical results where images and text are used as transmitted data, in Sect. 4.5 we present an analysis of the security provided by the algorithm as well as its ability to withstand known and chosen plaintext attacks, finally a conclusion and comments are given in Sect. 4.6.

4.2   Transmitter

The next definitions will be necessary for further understanding the properties of the transmitter and the development of the receivers:

Definition 4.1 Algebraic observability condition: A state variable is said to be algebraically observable in relation to the inputs and outputs of the system if it satisfies a differential polynomial equation in terms of the inputs, outputs and some of its time derivatives.

There is a class of chaotic systems that even if they does not fulfill the algebraic observability condition their states can be reconstructed in other ways, hence the following definition:

Definition 4.2 A system is said to be a Liouvillian system if its state variables that do not fulfill the algebraic observability condition, in turn, can be written as an expression in terms of integrals or exponentials of integrals of the output and some of its time derivatives.

For this encryption scheme the transmitter must be a Liouvillian chaotic system, the Colpitts oscillator is selected because it is a Liouvillian system, but also it has other desirable properties for secure communications such as stable and bounded dynamics which vary greatly depending on its initial conditions, this chaotic dynamic makes the masking of the transmitted data more effective since it is less likely to be unmasked without the proper receiver.

The Colpitts oscillator is a chaotic non-linear system that complies with the requirements to be a Liouvillian system. it has been used as a source for periodical signals (sinusoidal signals usually) and thanks to its chaotic behavior at low and high fundamental frequencies its use on secure communications is more feasible. Its dynamics are given by:

$$\dot{x}_1 = x_2 - f(x_3)$$
$$\dot{x}_2 = -x_1 - bx_2 - x_3 \tag{4.1}$$
$$\dot{x}_3 = x_2 - d$$
$$y = x_2$$

where:

$$f(x_3) = \begin{cases} -a(x_3 + 1) & x_3 < -1 \\ 0 & x_3 \geq -1 \end{cases}$$

The Colpitts oscillator based transmitter can be rewritten as follows:

$$x_1 = -\dot{y} - by - \int (y - d)$$
$$x_2 = y \tag{4.2}$$
$$x_3 = \int (y - d)$$

Then it is a Liouvillian system. Both transmitters encrypt the messages via chaotic masking, the first transmitter is intended to be used by the Super-Twisting observer and it masks data by embedding it into the state of the oscillator, this encryption scheme needs that the oscillator be written in a canonical form by means of a change of variable:

$$n_1 = x_3$$

$$n_2 = x_2 - d$$

$$n_3 = \gamma - x_1 - bx_2 - x_3$$

And its dynamic equations are:

$$\dot{n}_1 = n_2$$

$$\dot{n}_2 = n_3 + s$$

$$\dot{n}_3 = -2n_2 + d + f(n_1) - bn_3 \qquad (4.3)$$

$$y = n_2$$

The variable s is the data carrier signal to be masked. The second receiver will not require the transformation, and the encryption method will also mask the carrier signal with the state of the transmitter:

$$\dot{x}_1 = x_2 - f(x_3)$$

$$\dot{x}_2 = -x_1 - bx_2 - x_3 + s$$

$$\dot{x}_3 = x_2 - d \qquad (4.4)$$

$$y = x_2$$

This encryption scheme allows to avoid masking the message with the output of the system, this is because masking information that way makes the system susceptible to known plaintext attacks and chosen plaintext attacks.

The data carrier signal s is formed by the plain image or plain text, in the case of RGB images they will be formed by three matrices of unsigned eight bit integer numbers, these are integer numbers from 0 to 255, text messages are a single vector made of the same unsigned integers.

Suppose that the RGB image is of size $i \times j$ The easiest way to create the data carrier signal is to make a vector v_{int} with $i \times j \times 3$ elements and fill it with each element of the matrix, then divide it by 255 to obtain a vector formed by real numbers between 0 and 1, this is:

$$v = v_{int}/255$$

The carrier signal is made by defining a starting time t_s of transmission and an end time t_e and the amplitude of the signal takes the values of the vector and the amplitude will have period

$$P = \frac{t_e - t_s}{3 \cdot i \cdot j}$$

This form of carrier signal is very susceptible to reconstruction error as the slightest variation in the reconstruction of the signal will cause errors in the final recovered image, so it is not preferred by the authors, instead the integer vector elements are turned into their binary equivalent producing a new vector v_{bin} of size $i \times j \times 24$ composed of 0 and 1, then, the carrier signals amplitude is given by:

$$s = a \left(v_{bin} - 0.5 \right)$$

Where a is the amplitude required by the encryption scheme. This allows to register a negative number as 0 and a positive as 1, thus considerably reducing the effects of state reconstruction inaccuracy, of course, this comes at the price of having a signal eight times longer than the previous method. This last method is preferred by the authors for providing accurate image and text reconstructions.

4.3 Receiver

The design of the receiver will be specific to each transmitter, both use data masking as means of encryption, but each one will have its unique characteristics, the first one is the Super-Twisting observer

4.3.1 Super-Twisting Based Receiver

The receiver dynamic is given by the following equations:

$$\dot{\hat{n}}_1 = \hat{n}_2 + m\tau^{-1} k_c sign \left(y - \hat{y} \right)$$

$$\dot{\hat{n}}_2 = \hat{n}_3 + k_a \left(y - \hat{y} \right) + m\tau^{-1} \left| y - \hat{y} \right|^{\frac{1}{2}} sign \left(y - \hat{y} \right)$$

$$\dot{\hat{n}}_3 = k_b \left(y - \hat{y} \right) + m^2 \tau^{-2} sign \left(y - \hat{y} \right) \qquad (4.5)$$

$$\hat{y} = \hat{n}_2$$

$$\hat{s} = \hat{n}_2 - \hat{n}_3$$

Where $k_a > 0, k_b > 0, k_c > 0, m > 0$ y $0 < \tau < 1$ are positive constants and $\hat{s}$ is the recovered message given by:

$$\hat{s} = n_3 + s - \hat{n}_3$$

The message recovery error is:

$$e_s = s - \hat{s} = s - (n_3 + s - \hat{n}_3) = -n_3 + \hat{n}_3 \tag{4.6}$$

This error is bounded by the synchronization error:

$$\left\| n_3 - \hat{n}_3 \right\| \leq \left\| Cn - C\hat{n} \right\| = \| Ce \|$$

As the estimated state $\hat{n}$ converges to the transmitters state n, the reconstructed message $\hat{s}$ converges to the message. For proving the convergence of the messages the following assumptions are needed:

Assumption 4.1 *There exists non-negative constants L_{0f} y L_{1f} such that the following quasi-Lipschitz condition holds:*

$$\left\| \Delta_f \right\| \leq L_{0f} + \| e \| \left(L_{1f} + \| A_\mu \| \right)$$

Assumption 4.2 *The message is bounded by a positive real constant M:*

$$\| s \| \leq M$$

Assumption 4.3 *There is a positive definite matrix $0 < Q = Q^T$ such that the Riccati equation:*

$$P A_\mu + A_\mu^T P + PRP + Q = 0 \tag{4.7}$$

Has solution $0 < P = P^T$ with:

$$R = \Lambda_f^{-1} + 2 \left\| \Lambda_f \right\| L_{1f} I, \ 0 < \Lambda_f = \Lambda_f^T$$

$$Q = Q_0 + 2 \left(L_{1f} + \| A_\mu \| \right)^2 I$$

Assumption 4.4 *There is a vector*

$$K = m\tau^{-1} \begin{pmatrix} k_c \\ |e|^{\frac{1}{2}} \\ m\tau^{-1} \end{pmatrix} P^{-1} C^T \geq 0$$

where not all its components are zero.

4.3.2 Proof of Stability

The synchronization error is defined as:

$$e = \begin{bmatrix} e_1 \\ e_2 \\ e_3 \end{bmatrix} = \begin{bmatrix} n_1 - \hat{n}_1 \\ n_2 - \hat{n}_2 \\ \frac{1}{m}(n_3 - \hat{n}_3) \end{bmatrix}$$

The derivative of the error with an added real constant $\mu > 0$ is:

$$\begin{bmatrix} \dot{e}_1 \\ \dot{e}_2 \\ \dot{e}_3 \end{bmatrix} = \begin{bmatrix} e_2 + \mu e_1 - \mu e_1 - m\tau^{-1} k_c \text{sign}(y - \hat{y}) \\ e_3 - k_a(e) - m\tau^{-1} |e|^{\frac{1}{2}} \text{sign}(e) + \mu e_2 - \mu e_2 + s \\ \frac{1}{m}\Phi(n_1, n_2, n_3) - \frac{k_b}{m}(e) - m^2\tau^{-2}\text{sign}(e) + \mu e_3 - \mu e_3 \end{bmatrix}$$

With $\Phi(n_1, n_2, n_3) = -2n_2 + d + f(n_1) - bn_3$, this last equality can be written as follows:

$$\dot{e} = \begin{bmatrix} -\mu & 1 & 0 \\ 0 & -\mu & 1 \\ 0 & 0 & -\mu \end{bmatrix} e - \begin{pmatrix} 0 \\ k_a \\ k_b \end{pmatrix} \begin{bmatrix} 0 & 1 & 0 \end{bmatrix} \begin{bmatrix} e_1 \\ e_2 \\ e_3 \end{bmatrix}$$

$$-m\tau^{-1} \begin{pmatrix} k_c \\ |e|^{\frac{1}{2}} \\ m\tau^{-1} \end{pmatrix} \text{sign} \left(\begin{bmatrix} 0 & 1 & 0 \end{bmatrix} \begin{bmatrix} e_1 \\ e_2 \\ e_3 \end{bmatrix} \right) + \begin{pmatrix} \mu e_1 \\ \mu e_2 + s \\ \frac{\Phi(n_1, n_2, n_3)}{m} + \mu e_3 \end{pmatrix}$$

The derivative of the error can be written as:

$$\dot{e} = A_\mu e - k_2 Ce - k_1 \text{sign}(Ce) + \Delta_f \tag{4.8}$$

With:

$$k_1 = -m\tau^{-1} \begin{pmatrix} k_c \\ |e|^{\frac{1}{2}} \\ m\tau^{-1} \end{pmatrix}, k_2 = \begin{pmatrix} 0 \\ k_a \\ k_b \end{pmatrix}, \Delta_f = \begin{pmatrix} \mu e_1 \\ \mu e_2 + s \\ \frac{\Phi(n_1, n_2, n_3)}{m} + \mu e_3 \end{pmatrix}$$

The gain $k_2 > 0$ is chosen so the effects of the message within the uncertainty Δ_f are reduced. The receiver allows for the error $e_s = s - \hat{s}$ to remain bounded and for it to converge to the residual set:

$$D_\varepsilon = \{e_s| \, \|e_s\|_P \leq \bar{\mu}(k)\}$$

Where P is the solution to the Riccati equation, then:

$$\bar{\mu}(k) = \left(\frac{\rho(K)}{\sqrt{\left(K\alpha_p\right)^2 + \rho(K)\alpha_Q + K\alpha_p}} \right)$$

Where:

$$\rho(K) = 2 \, \|\Lambda_f\| \, L_{0f}^2 + 4k_2 s^+ \sqrt{n\Lambda_f^{-1}}$$

$$K\alpha_p = K \left(\lambda_{min} \left(P^{-1/2} C^T C P^{-1/2} \right) \right)$$

$$\alpha_Q = \lambda_{min} \left(P^{-1/2} Q^T Q P^{-1/2} \right)$$

For proving this the next Lyapunov candidate function is proposed:

$$V(e) = \|e\|_P^2 = e^T P e, \; 0 < P = P^T$$

Its derivative is:

$$\dot{V}(e) = 2e^T P A_\mu e - 2e^T P k_2 C e - 2e^T P k_1 sign(Ce) + 2e^T P \Delta f$$

The gains are chosen $k_a > 0$ y $k_b > 0$ and in view of Assumption 4.4

$$\dot{V}(e) \leq 2e^T P A_\mu e - 2K e^T C^T sign(Ce) + 2e^T P \Delta f$$

Using the inequality $X^T Y + Y^T X \leq X^T \Lambda_f X + Y^T \Lambda_f^{-1} Y$ for $0 < \Lambda_f = \Lambda_f^T$:

$$\dot{V}(e) \leq e^T (P A_\mu + A_\mu^T P) e - 2K e^T C^T sign(Ce) + e^T P \Lambda^{-1} P e + \Delta_f^T \Lambda_f \Delta_f$$

From Assumption 4.1:

$$\dot{V}(e) \le e^T(PA_\mu + A_\mu^T P + PRP + Q)e - e^T Qe + L_{0f}^2$$

$$+ 2\|\Lambda_f\|\left[\|e\|^2\left(L_{1f} + \|A_\mu\|^2\right)\right] - 2e^T C^T K sign(Ce)$$

$$\dot{V}(e) \le e^T(PA_\mu + A_\mu^T P + PRP + Q)e - e^T Qe$$

$$+ 2\|\Lambda_f\| L_{0f}^2 - 2e^T C^T K sign(Ce)$$

By means of Assumption 4.3:

$$\dot{V}(e) \le -e^T Qe + 2\|\Lambda_f\| L_{0f}^2 - 2K\sum_{i=1}^{n}|(Ce)_i|$$

$$\dot{V}(e) \le -e^T Qe - 2K\sum_{i=1}^{n}|(Ce)_i| + \rho(k) \tag{4.9}$$

With $\rho(k) = 2\|\Lambda_f\| L_{0f}^2 + 4k\bar{s}\sqrt{n\Lambda_f^{-1}}$, then:

$$\dot{V}(e) \le -\|e\|_Q - 2K\alpha_P \|e\|_P + \rho(k)$$

Then:

$$\left(\sum_{i=1}^{n}|(Ce)_i|\right)^2 \ge \sum_{i=1}^{n}|(Ce)_i|^2 = \|Ce\|^2 = \left\|CP^{-1/2}P^{-1/2}e\right\|^2 \ge \alpha_P e^T Qe$$

Where $\alpha_P = \lambda_{min}\left(P^{-1/2}C^T CP^{-1/2}\right)$, so:

$$\dot{V}(e) = \frac{d}{dt}\|e\|_P^2 \le -\|e\|_Q^2 - 2K\alpha\|e\|_P + \rho(k)$$

It yields to:

$$\dot{V}(e) = -\alpha_Q V(e) - \vartheta\sqrt{V(e)} + \beta \tag{4.10}$$

Having $\alpha_Q = \lambda_{min}\left(P^{-1/2}Q^T QP^{-1/2}\right) > 0$, $\vartheta = 2K\alpha_p$ y $\beta = \rho(k)$. From Assumptions 4.1 and 4.3, it is possible to conclude that:

$$\left[1 - \frac{\bar{\mu}(k)}{V(e)}\right]_+ \to 0$$

Where the function $[\bullet]_+$ is defined as:

$$[z]_+ = \begin{cases} z & , z \geq 0 \\ 0 & , z < 0 \end{cases}$$

So $V(e) \leq \bar{\mu}(k)$, then:

$$V(e) = \|e\|_p = e^T P e \geq e_2^T P e_2$$

$$e_1 = Cn - C\hat{n} = -e_s$$

$$e_1^T P e_1 = \left(-e_s^T\right) P (-e_s) = e_s^T P e_s = \|e_s\|_P$$

The message estimation error converges to the set D_ε since:

$$\|e_s\|_P = \|e_2\|_P \leq \bar{\mu}(k)$$

So the message recovery error converges to $\bar{\mu}(k)$:

$$\|e_s\|_P \leq \bar{\mu}(k)$$

The synchronization error remains bounded even with the presence of the message embedded into the state and as consequence the message recovery error remains bounded showing the main characteristics of this receiver.   □

For recovering the message it is necessary to have access to the derivative of the output, but directly applying a derivative to this kind of signals has detrimental consequences to the quality of the recovered information as the derivative itself tends to amplify noise, an alternative to this is to approximate the derivative by means of:

$$\dot{n}(t) \approx \frac{n(t) - n(t - h)}{h}$$

The last equation returns the slope of the line between the current state $n(t)$ and a previous measurement of the state $n(t - h)$ taken h seconds before. This approximation will not amplify any noise thus providing less error than applying a derivative to signal.

This receiver does not require knowledge of the transmitter dynamical equations, then, they remain hidden to the end users providing some degree of safety to the data, aside from the one given by the chaotic masking, but still, this encryption scheme depends mostly on the chaotic masking, this is because it is not possible to produce encryption keys that depend on the initial conditions of the transmitter or the gains of the receiver because, usually, observer states will converge to the transmitters states regardless of small variations on its initial conditions or gains. Another setback of sliding modes based observers is that chattering does not allow the error to remain close to zero for extended periods of time, then error will always be present in the reconstruction of the transmitted message.

The next receiver does not have the problems caused by using observers as receiver, and it will also retain the characteristic of not providing the end user with full knowledge of the transmitters dynamical equations.

4.3.3 Reconstruction of the States Based Receiver

The design of this receiver is based on the property that makes a system be Liouvillian (see Definition 4.1), this property allows to reconstruct the states of a system without the need of an state observer, so it is possible to create receivers that will not be affected by the limitations of observers, and yet they will retain the ability of not using the same dynamic equations of the transmitter. The message will be masked within the dynamic equations of the system and recovering it will require to reconstruct the dynamical equation of the output state, this state can be reconstructed with the output of the transmitter so the message can be recovered. The receiver is given by:

$$\hat{x}_1 = \dot{y} - by - \int (y - d))$$

$$\hat{x}_2 = y$$

$$\hat{x}_3 = \int (y - d)$$

$$\hat{s} = \dot{y} + \hat{x}_1 + b\hat{x}_2 + b\hat{x}_3$$

The reconstruction of the message requires an accurate reconstruction of the state, since the system is Liouvillian, the other two states can accurately be reconstructed by integrating the output and its derivatives, so error will not be present in this reconstruction:

$$e = \begin{bmatrix} x_1 - \hat{x}_1 \\ x_2 - \hat{x}_2 \\ x_3 - \hat{x}_3 \end{bmatrix}$$

$$\begin{bmatrix} x_1 - \hat{x}_1 \\ x_2 - \hat{x}_2 \\ x_3 - \hat{x}_3 \end{bmatrix} = \begin{bmatrix} \int \left(y - f \left[\int (y - d) dt \right] \right) dt - \int \left(y - f \left[\int (y - d) dt \right] \right) dt \\ y - y \\ \int (y - d) dt - \int (y - d) dt \end{bmatrix}$$

The synchronization error is zero:

$$\begin{bmatrix} x_1 - \hat{x}_1 \\ x_2 - \hat{x}_2 \\ x_3 - \hat{x}_3 \end{bmatrix} = \begin{bmatrix} 0 \\ 0 \\ 0 \end{bmatrix}$$

$$e = 0$$

Then the message recovery error is defined as:

$$e_s = s - \hat{s} = s - \left(\dot{y} + \hat{x}_1 + b\hat{x}_2 + b\hat{x}_3 \right)$$

From (4.4):

$$s = \dot{x}_2 + x_1 + bx_2 + x_3$$

Considering that $e = 0$

$$e_s = 0$$

The derivative of the output is approximated by:

$$\dot{y}(t) \approx \frac{y(t) - y(t-h)}{h}$$

This receiver is very sensitive to initial conditions, if they are not equal, the message recovery error will be different to zero making impossible to have access to the encrypted data, this feature can be used to produce an encryption key formed by the set of initial condition and parameters that produce chaotic behavior in the transmitter.

The combination of the transmitter and the receiver give better security features to the transmitted data than the ones provided by state observers, because it is possible to use encryption keys and the amplitude of the data carrier signal can be made arbitrarily small and as in the case of the Super-Twisting observer, it does not use the full dynamic of the transmitter nor the same equations.

4.4   Numerical Simulation

For testing the results given in previous sections image and text are encrypted and recovered by the observer and the reconstruction based receiver, for this the parameters of the Super-Twisting observer are:

Parameter	Value
k_1	10
k_2	15
k_3	20
k_4	25

The parameters of the transmitter and Liouvillian receiver are:

Parameter	Value
a	2
b	3
d	0.6

The maximum amplitude of the data carrier signal will be of 3, this value is chosen so the derivative approximation does not affects the recovery of the data. The transmitted image is the presented in Fig. 4.1. The encrypted images are shown in Figs. 4.2 and 4.3.

In both cases the encrypted image does not present any data of the plain image, then the masking is effective, because even if the amplitude of the data carrier signal is almost

Fig. 4.1 Transmitted data

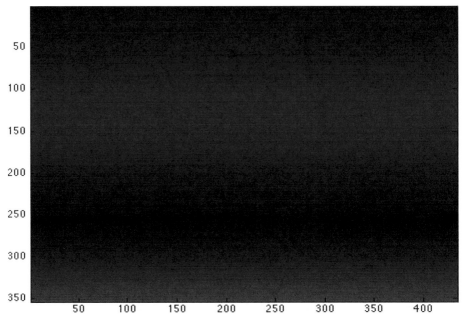

Fig. 4.2 Encrypted data by Super-Twisting observer

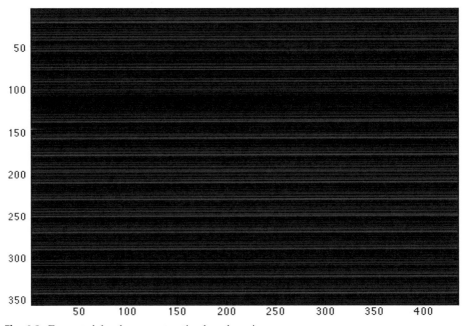

Fig. 4.3 Encrypted data by reconstruction based receiver

Fig. 4.4 Recovered data by Super-Twisting observer

the same of the states, the plain image is not visible in the cypher image. Although the masking is equally effective in both transmitters, the quality of the reconstructed images is completely different, in Figs. 4.4 and 4.5 it is possible to see the clear advantage of the Liouvillian transmitter.

The recovered image is more accurate in the reconstruction based receiver (Fig. 4.5), this mainly because the observer is intended to bound the error in the tracking of the transmitter states, yet the recovered data is still useful and the transmitted image can be seen but not as clearly as in Fig. 4.4. For text transmission a fragment of the abstract is used:

Transmitted text:
In this chapter we show the use of a Liouvillian chaotic system on a master-slave scheme for secure communications. Two forms of data encryption are presented, the first is a Super-Twisting state observer based on the Super-Twisting algorithm

Fig. 4.5 Recovered data by reconstruction based receiver

Recovered text by Super-Twisting observer:
IIn this chapter we show the use of a Liouvillian chaotic system on a master+slave scheme for secure communications. Two forms of data encryption are preseeted, the first is a Super-Twisting state observer based onnthe Super-Twistimf'kfnqhsgl+sgdrdbnmchr'

Recovered text by reconstruction receiver
In this chapter we show the use of a Liouvillian chaotic system on a master-slave scheme for secure communications. Two forms of data encryption are presented, the first is a Super-Twisting state observer based on the Super-Twisting algorithm

As in the image data, the observer produces error in the recovered data in comparison to the reconstruction based receiver, this last receiver produces an accurate reconstruction message as it has no error in any of the recovered characters.

4.5 Vulnerability to Cryptanalysis

Usually chaotic masking is done by combining the output of the transmitter with the message carrier signal [3, 14, 16], this type of masking is given by the next equation:

$$y = Cx + s$$

Masking data in this way makes the cryptosystem extremely vulnerable to known and chosen plaintext attacks, Since the signal Cx depends entirely on the key, then, if the key is the same for each message, Cx will always be the same and only s will change. In a known plaintext attack, the opponent has access to a number of plaintext s and ciphertext y pairs, in the case of this masking if the opponent has a single pair, the entire system is compromised because they will be able to reconstruct Cx:

$$Cx = y - s$$

In consequence any another message s_1 that is transmitted with the same key will have the same vector Cx masking s_1, the ciphertext of this new message will be:

$$y_1 = Cx + s_1$$

Recovering s_1 without the need of the receiver nor the key is simple as the vector Cx is already known:

$$s_1 = y_1 - Cx$$

Then this common chaotic masking will be vulnerable to known plaintext attacks. In turn, the proposed masking based on the Liouvillian property is not vulnerable to this type of attack, as the cipher text or image depends not only on the states but also depends on the message, so each state vector Cx will be unique to each message, because the encryption is given by the equation:

$$\dot{x} = f(x, s)$$
$$y = Cx$$

Making very difficult to implement this type of known plaintext attacks due to Cx not being the same for each message. To further exemplify the encryption algorithm ability to withstand known plaintext attacks we will implement a chosen plaintext attack, this type of attack is far more dangerous than the known plaintext attack. In the chosen plaintext attack, the attacker is allowed to choose the message that will be transmitted, the messages

Fig. 4.6 Chosen image

are designed so they could lead to exposing a weakness in the algorithm, recover another unknown message or in the worst case to recover the encryption key.

For this example we choose as message a completely black image (formed only by zeros) with the same size of the image used in the previous sections with the purpose of recovering the tiger image without the need of the key, this by retrieving the key stream (Cx) unaffected by the image, consider that the key will be the same for both messages. The obtained results are shown in Figs. 4.6 and 4.7.

As can be seen in Figs. 4.6 and 4.7, the chosen plaintext attack fails to recover the encrypted image or any data that could compromise the key, this is caused by the message itself, as the key stream depends on both the message and the key, then every different message will produce a different key stream even if the key is not changed, so this allows to use a single key multiple times, also it is worth to mention that this is not a common feature in this type of encryption schemes.

4.6   Concluding Remarks

Both encryption methods where able to produce recovered messages, but as expected the reconstruction based receiver made possible to recover the data without significant error, this last method retained the security feature of the Super-Twisting observer which is not to give the user of the receiver the same dynamical equations of the transmitter oscillator, even more it considerably enhanced the security of the transmitted data, this because the encryption key based on the initial conditions and the parameters of the

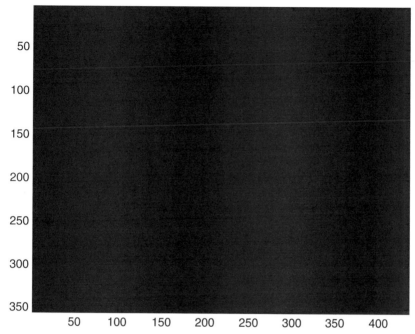

Fig. 4.7 Recovered image by chosen plaintext attack

system, its biggest advantage over most state observers is that it makes possible to use encryption keys, not leaving the security of the message to relay solely on the chaotic masking. Another advantage of the reconstruction based receiver is that it is capable of reconstructing the transmitted data without error, this can be seen in Figs. 4.4 and 4.5, in this, the image recovered by the Super-Twisting observer presents inaccuracy in the color, this is caused by the design of the observer, since it is intended to bound the error only, it will present small differences in the reconstructed signals, then, this error is carried over to the reconstructed data and produces the effect seen in Fig. 4.4. In comparison the performance of the reconstruction based receiver is better, as the recovered data is identical to the transmitted data. It is possible to conclude that the properties that make a system be Liouvillian can be very convenient in data encryption, as explained, they present advantages in security and quality of the recovered information plus reducing the vulnerability to known plaintext attacks.

References

1. Chua, L. O., Kocarev, L., Eckert, K., & Itoh, M. (1992). Experimental chaos synchronization in Chua's circuit. *International Journal of Bifurcation and Chaos, 2*(3), 705–708.
2. Morgül, Ö., & Solak, E. (1996). Observer based synchronization of chaotic systems. *Physical Review E, 54*(5), 4803.

3. Hassan, M. F. (2014). Observer design for constrained nonlinear systems with application to secure communication. *Journal of the Franklin Institute, 351*(2), 1001–1026.

4. Alvarez, G., & Li, S. (2006). Some basic cryptographic requirements for chaos-based cryptosystems. *International Journal of Bifurcation and Chaos, 16*(8), 2129–2151.

5. Volos, Ch. K., Kyprianidis, I. M., & Stouboulos, I. N. (2013). Image encryption process based on chaotic synchronization phenomena. *Signal Processing, 93*(5), 1328–1340.

6. Roohbakhsh, D., & Yaghoobi, M. (2015). Fast adaptive image encryption using chaos by dynamic state variables selection. *International Journal of Computer Applications, 113*(12), 28–32.

7. Schmitz, R. (2001). Use of chaotic dynamical systems in cryptography. *Journal of the Franklin Institute, 338*(4), 429–441.

8. Akhavan, A., Samsudin, A., & Akhshani, A. (2011). A symmetric image encryption scheme based on combination of nonlinear chaotic maps. *Journal of the Franklin Institute, 348*(8), 1797–1813.

9. Volos, C. K., Kyprianidis, I. M., Stouboulos, I., & Pham, V. T. (2015). Image encryption scheme based on non-autonomous chaotic systems. In *Computation, cryptography, and network security* (pp. 591–612). Springer International Publishing.

10. Martínez-Guerra, R., Cruz-Victoria, J., Gonzalez-Galan, R., & Aguilar-Lopez, R. (2006). A new reduced-order observer design for the synchronization of lorenz systems. *Chaos, Solitons & Fractals, 28*(2), 511–517.

11. Martínez-Guerra, R., Gómez-Cortés, G. C., & Pérez-Pinacho, C. A. (2015). *Synchronization of integral and fractional order chaotic systems a differential algebraic and differential geometric approach with selected applications in real-time.* Springer.

12. Sobhy, M. I., & Shehata, A. E. R. (2001). Chaotic algorithms for data encryption. In *IEEE International Conference on Acoustics, Speech, and Signal Processing (ICASSP '01)* (Vol. 2, pp. 997–1000).

13. Dachselt, F., & Schwarz, W. (2001). Chaos and cryptography. *IEEE Transactions on Circuits and Systems I: Fundamental Theory and Applications, 12*(48), 1498–1509.

14. Cheng, C. C., Lin, Y. S., & Wu, S. W. (2012). Design of adaptive sliding mode tracking controllers for chaotic synchronization and application to secure communications. *Journal of the Franklin Institute, 349*(8), 2626–2649.

15. Levant, A. (1993). Sliding order and sliding accuracy in sliding mode control. *International Journal of Control, 58*(6), 1247–1263.

16. Castro-Ramírez, J., Martínez-Guerra, R., & Cruz-Victoria, J. C. (2015). A new reduced-order observer for the synchronization of nonlinear chaotic systems: An application to secure communications. *Chaos: An Interdisciplinary Journal of Nonlinear Science, 25*(10), 103128.

State Observers and Cryptography

5

Abstract

In this chapter we propose the use of chaotic systems on a master-slave scheme for secure communications. This cryptosystem allows to use the advantages of block cyphers in color images, these features permit us to encrypt messages as large as images without the need of keys bigger than the transmitted data, this is achieved by using the synchronization of two chaotic oscillators and fractals such as Julia sets. Two different receivers will be presented; the first one is an exponential polynomial observer and the second is a reconstruction of the states based on the properties of Liouvillian systems, the latter retaining the same security features but with a more simple implementation and faster encryption and decryption times.

5.1   Introduction

Synchronization of chaotic systems has received a lot of attention from researchers due to its potential applications [1] in special there have been many advances in secure communications [2–4] where the goal is to encrypt data using signals provided by chaotic oscillators, in this application of synchronization one of the most common solutions is to design the receiver as a state observer in order to achieve synchronization to a nonlinear electric circuit and recover the information that was used to encrypt the transmitted data and with this information recover the message [5, 6]. The proposed encryption algorithm uses the synchronization of chaotic systems under master-slave paradigm, where the transmitter is named the master system and the receiver is called the slave system, the intention of this configuration is to allow the slave to reconstruct the states of the master

© The Author(s), under exclusive license to Springer Nature Switzerland AG 2023
R. Martínez-Guerra et al., *Encryption and Decryption Algorithms for Plain Text and Images using Fractional Calculus*, Synthesis Lectures on Engineering, Science, and Technology, https://doi.org/10.1007/978-3-031-20698-6_5

system by using a signal provided by the latter, then with this reconstruction, the encrypted data is recovered.

Various encryption algorithms rely on shuffling and diffusing the message data, these algorithms are quite susceptible to chosen plaintext attacks or known plaintext attacks regardless of how well the distortion of the original data is done. In recent developments algorithms that are more resistant to this type of attacks have been introduced, this characteristic has been achieved by making the shuffling and diffusing operations depend on the key and the transmitted message itself, but in order to accomplish this operation, they require to provide the user with more information than the key and ciphertext or image and will also have a slight error caused by the extra data needed to recover the message.

The proposed encryption algorithm will have these security features, but will not require to give more data to the user, nor will it sacrifice a portion of the message to hide this data, this is achieved through the properties of Liouvillian systems and the use of Julia sets for reducing the relation between the encrypted data and the key. Two approaches for data recovery will be presented, the first is the exponential polynomial observer (EPO), this observer can reconstruct all the states of the transmitter and thus recover the encrypted data. The second receiver is a reconstruction of the states that are involved in data encryption based on the properties of Liouvillian systems, the reconstruction will also solve the lack of accuracy in data reconstruction caused by the observer estimation error. The two encryption methods will be tested with color images as transmitted data, making possible to present the advantages of the use of Liouvillian systems in secure communications.

The rest of the chapter is organized as follows. In Sect. 5.2, some definitions and the proposed algorithm encryption is introduced. In Sect. 5.3, the method for data recovery is presented with stability analysis for the exponential polynomial observer as well as the Liouvillian property of the system is used for message recovery. In Sect. 5.4, numerical simulation of the proposed methodology is presented in order to analyse the algorithm. Finally, in Sect. 5.5, some concluding remarks are given.

5.2 Encryption

5.2.1 Generating Pseudo-Random Numbers

The class of systems that will be introduced in the following definition will the main focus of this chapter.

Definition 5.1 A system is called Liouvillian if its state variables can be written as an expression in terms of integrals or exponentials of integrals of the output and some of its time derivatives.

Let us consider the following example.

Example 5.1 Let the dynamic equations that describe the Colpitts oscillator given by:

$$\dot{x}_1 = x_2 - f(x_3)$$
$$\dot{x}_2 = -x_1 - bx_2 - x_3$$
$$\dot{x}_3 = x_2 - d \tag{5.1}$$
$$y = x_2$$

where:

$$f(x_3) = \begin{cases} -a(x_3 + 1) & x_3 < -1 \\ 0 & x_3 \geq -1 \end{cases}$$

The states of the Colpitts oscillator can be written as follows:

$$x_1 = -\dot{y} - by - \int (y - d)$$
$$x_2 = y \tag{5.2}$$
$$x_3 = \int (y - d)$$

Showing that is a Liouvillian system.

For secure communications, it is necessary to generate a sequence of numbers, chaotic oscillators are a good source of pseudorandom numbers, in this case the Colpitts oscillator will be used because it is a Liouvillian system, but also it has other desirable properties for encryption like a stable and bounded dynamics, that changes greatly depending on its initial conditions.

5.2.2 Encryption Algorithm

The encryption process consist in diffusion and shuffling of the plain image, this stages will depend on the key and the message causing that known and chosen plaintext attacks are difficult to implement. In the description of the algorithm the message will be assumed to be an RGB color image, this algorithm is not limited to images and can also be applied to text messages. Consider the RGB image given by $P \in \mathbb{Z}^{m \times n \times 3}$ composed by three matrices P_r, P_g and P_b of size $m \times n$.

The key is a set of hexadecimal characters, long enough to obtain the information required by the encryption process, the key will be composed as is shown below:

$$K_{ey} = K_1 - K_2 - K_3 - K_4 - \cdots - K_l \cdots$$

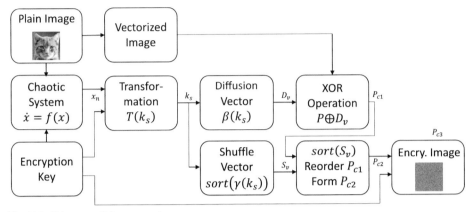

Fig. 5.1 Diagram of the encryption process

Each key segment will be identified as: $K_i = ABCD$, with $1 \leqslant i \leqslant l$. Every segment of the key contains information about the process such as initial conditions, parameters, sample times or gains. Here we give the following steps (Fig. 5.1):

Step 1 Create the initial conditions for the chaotic oscillator with the plain image. By using a three state oscillator, each state can be used for each one of the RGB matrices that compose the image. The initial condition for the first state is generated as follows:

$$xo_1 = \frac{\sum_{i=1}^{m} \sum_{j=1}^{n} P(i, j, 1)}{255mn} \tag{5.3}$$

Then $xo_1 \in (0, 1)$ and can be used as initial condition for the state x_1. The process must be repeated for the other two states using the remaining matrices.

Step 2 With the first sections of the key generate the chaotic oscillators parameters. The Colpitts oscillator has three parameters, the first one can be formed by $K_1 = ABCD$, the parameter a is:

$$a = \frac{ABCD}{FFFF}$$

This will provide $a \in (0, 1)$, the remaining element of the key acts as scale factor for the parameter if the oscillator needs it, this step must be repeated for the remaining oscillators parameters b and d.

Step 3 Use the chaotic oscillator to generate a long trajectory. Once the oscillator parameters and initial conditions have been obtained, it must generate sufficiently long trajectories for them to provide enough random numbers to make vectors for the shuffling and diffusion operations.

Step 4 With the second section of the key K_2 obtain numbers from the oscillators state trajectories. To do this, create a matrix k_s based on the transmitter oscillator's states x_n and a sample time t specified in the second section of the key, the sample time is obtained by:

$$t = \left(\frac{ABCD}{FFFF}\right) 10^{1-E} \tag{5.4}$$

and the matrix is:

$$k_s = \begin{bmatrix} x_1\,(t) & x_2\,(t) & \cdots & x_n\,(t) \\ x_1\,(2t) & x_2\,(2t) & \cdots & x_n\,(2t) \\ \vdots & \vdots & \ddots & \vdots \\ x_1\,(nt) & x_2\,(nt) & \cdots & x_n\,(nt) \end{bmatrix} \tag{5.5}$$

Then the matrix k_s must be modified by a non-invertible transformation, so the relation between the diffusion and shuffling stages and the states of the oscillator is reduced due to the fact that the Julia set is not an injective function, this transformation will be made through the Julia set [3]:

$$Z_{n+1} = Z_n^2 + b, \quad n \in [1, \alpha], \quad b \in \mathbb{C} \tag{5.6}$$

$$b = c + di$$

$$c = \frac{ABCDE}{2FFFFF}$$

$$d = \frac{ABCDE}{2FFFFF} \tag{5.7}$$

$$i = \sqrt{-1}$$

The key will provide the value of b and the number of iterations of the Julia set $\alpha = ABCDE$, choosing $b = c + di$ guarantees that the values provided by the fractal are always within the attractor. The Julia set requiring complex numbers will make possible to further modify the numbers given by the Liouvillian system and making very hard to find out which numbers produced the order of shuffling and values for diffusion, the initial condition will be made by the value of $x_n\,(t_n)$ and another element of the key:

$$Z_0 = \left[0.5 - \frac{x_n\,(t_n)}{ABCDE}\right] + \left[\frac{x_n\,(t_n)}{ABCDE}\right] i$$

Finally the transformation will be the norm of the complex value given by the Julia set:

$$T(x_n) = |Z_n|$$

Then, the matrix k_s is transformed into the matrix k_{st} by applying the Julia set transformation to each one of its elements:

$$k_{st} = T(k_s) = \begin{bmatrix} Z_{1,t_1} & Z_{2,t_1} & \cdots & Z_{n,t_1} \\ Z_{1,t_2} & Z_{2,t_2} & \cdots & Z_{n,t_2} \\ \vdots & \vdots & \ddots & \vdots \\ Z_{1,t_n} & Z_{2,t_n} & \cdots & Z_{n,t_n} \end{bmatrix} \tag{5.8}$$

The value t_n denotes the sample taken from state x_n at time $n \cdot t$.

Step 5 Create the shuffling order and the diffusion values from the previously generated numbers, the diffusion vector D_v is formed from k_{st}:

$$D_{v1} = \begin{bmatrix} Z_{1,t_1} & Z_{2,t_1} & \cdots & Z_{n,t_1} \end{bmatrix}$$

$$D_{v2} = \begin{bmatrix} Z_{1,t_2} & Z_{2,t_2} & \cdots & Z_{n,t_2} \end{bmatrix} \tag{5.9}$$

$$D_{v3} = \begin{bmatrix} Z_{1,t_n} & Z_{2,t_n} & \cdots & Z_{n,t_n} \end{bmatrix}$$

And combine it with the plain image by a bitwise XOR operation $\oplus$ to produce the partially encrypted image P_{c1}:

$$P_{cr1} = P_r \oplus D_{v1}$$

$$P_{cg1} = P_g \oplus D_{v2} \tag{5.10}$$

$$P_{cb1} = P_b \oplus D_{v3}$$

$$P_{c1} = P \oplus D_v \tag{5.11}$$

With the transformed vector k_{st} create a shuffle vector having as many elements as the message:

$$S_{v1} = \begin{bmatrix} Z_{1,t_1} & Z_{1,t_2} & \cdots & Z_{1,t_{mn}} \end{bmatrix}$$

$$S_{v2} = \begin{bmatrix} Z_{2,t_1} & Z_{2,t_2} & \cdots & Z_{2,t_{mn}} \end{bmatrix} \tag{5.12}$$

$$S_{v3} = \begin{bmatrix} Z_{3,t_1} & Z_{3,t_2} & \cdots & Z_{3,t_{mn}} \end{bmatrix}$$

$$S_v = sort(S_{v1}, S_{v2}, S_{v3}), \quad S_v \in \mathbb{R}^{mn} \tag{5.13}$$

Sort the elements of the shuffle vector and reorder the elements of the partially encrypted image P_{c1} according to the order of the elements of the shuffle vector to form P_{c2}.

$$P_{c2} = sort\,(P_{c1}, S_v) \qquad (5.14)$$

The encrypted image will be P_{c2}.

5.3   Data Recovery

To recover the encrypted message it will be necessary to reconstruct the trajectories of the Liouvillian oscillator, then by using the key it is possible to build the diffusion vector D_v and the shuffle vector S_v, the states of the oscillator will be reconstructed by a state observer and by a reconstruction of the states that use the properties of Liouvillian systems and to finally recover the encrypted image, the cypher image P_{c2} is reordered in function of S_v to produce P_{c1} and then the operation XOR is applied with D_v providing the plain image.

5.3.1   Exponential Polynomial Receiver

The exponential polynomial observer function is to estimate the states of a nonlinear system, it is similar to a Luenberger observer in the sense that it requires a reconstruction of the system dynamic and has a similar structure, it has more gain vectors and also has the characteristic that it will make the synchronization error of the states decrease more rapidly than an exponential bound, hence the name exponential polynomial. The dynamic of the observer is given by:

$$\dot{\hat{x}} = A\hat{x} + \psi\left(\hat{x}\right) + \sum_{i=1}^{m} K_i \left(y - C\hat{x}\right)^{2i-1}$$

$$\hat{y} = C\hat{x}$$

Where $\hat{x} \in \mathbb{R}^n$ are the states of the observer, y is the output of the transmitter, $\psi\left(\hat{x}\right)$ is the nonlinear part of the transmitter which satisfies the Lipschitz condition and $K_i \in \mathbb{R}^n$, $1 \le i \le m$ are the gain vectors of the observer. The following assumptions will be needed during the theoretical result of this section:

Assumption 5.1 *For a given $\varepsilon > 0$ and $A \in \mathbb{R}^{n \times n}$ there exists a matrix $P = P^T > 0$, $P \in \mathbb{R}^n$ that is the solution to the algebraic equation:*

$$A^T P + PA + L^2 P^2 + (1+\varepsilon) I = 0$$

Assumption 5.2 *The nonlinear part of the transmitter $\psi\left(\hat{x}\right)$ satisfies the condition:*

$$2\hat{x}^T P\psi\left(\hat{x}\right) \leq L^2\hat{x}^T P^2\hat{x} + \hat{x}^T\hat{x}$$

5.3.2 Stability

Proof The dynamic of the synchronization error is $\dot{e} = \dot{x} - \dot{\hat{x}}$, then:

$$\dot{e} = Ax + \psi\left(x\right) - \ldots$$

$$\ldots - \left(A\hat{x} + \psi\left(\hat{x}\right) + \sum_{i=1}^{m} K_i\left(y - C\hat{x}\right)^{2i-1}\right)$$

and considering that a single state variable will be used as output, then it is possible to make:

$$\dot{e} = A\left(e\right) + \phi(e) - \sum_{i=1}^{m} K_i Ce^{2i-1}$$

where $\phi(e) = \psi\left(x\right) - \psi\left(\hat{x}\right)$. From Assumption 5.2 the nonlinear part of the error $\phi(e)$ satisfies the condition:

$$2e^T P\phi(e) \leq L^2 e^T P^2 e + e^T e$$

The following Lyapunov candidate function is proposed:

$$V = e^T Pe$$

and then, by computing the derivative:

$$\dot{V} = \dot{e}^T Pe + e^T P\dot{e}$$

$$\dot{V} = \left[A\left(e\right) + \phi(e) - \sum_{i=1}^{m} K_i Ce^{2i-1}\right]^T Pe$$

$$+ e^T P\left[A\left(e\right) + \phi\left(e\right) - \sum_{i=1}^{m} K_i Ce^{2i-1}\right] \tag{5.15}$$

$$\dot{V} \leq e^T\left(A^T P + PA + L^2 P^2 + I\right)e$$

$$- 2e^T P\sum_{i=1}^{m} K_i Ce^{2i-1}$$

Considering that our interest lies in observers of order two or more ($m \geq 2$):

$$2e^T P \sum_{i=1}^{m} K_i C e^{2i-1} = 2e^T P K_1 C e + (Ce)^2 \, 2e^T P K_2 C e +$$

$$\cdots + (Ce)^{2m-2} \, 2e^T P K_m C e$$

Defining $M_1 = P K_1 C$, $M_2 = P K_2 C, \cdots$, $M_m = P K_m C \geq 0$ and since $e^T M_m e$ are scalar numbers $e^T M_m e = \left[e^T M_m e \right]^T$, then:

$$(Ce)^0 \, e^T M_1 e + (Ce)^1 \left(e^T M_1 e \right)^T$$

$$\cdots \quad + (Ce)^2 \, e^T M_2 e + (Ce)^2 \left(e^T M_1 e \right)^T$$

$$\cdots \quad (Ce)^{2m-2} \, e^T M_m e + (Ce)^{2m-2} \left(e^T M_m e \right)^T$$

$$= \sum_{i=1}^{m} (Ce)^{2i-2} \, e^T \left(M_i + M_i^T \right) e$$

The last equation shows that $2e^T P \sum_{i=1}^{m} K_i C e^{2i-1}$ is definite positive then, the only term that remains to be made definite negative is:

$$\dot{V} \leq e^T \left(A^T P + P A + L^2 P^2 + I \right) e$$

From the Assumption 5.1: $A^T P + P A + L^2 P^2 + I \leq -\varepsilon I$, consequently $\dot{V} \leq -\varepsilon \| e \|^2$. Consider that $V = \| e \|^2$ so $\alpha \| e \|_P^2 \leq V \leq \gamma \| e \|_P^2$, $\alpha = \lambda_{min}(P)$, $\gamma = \lambda_{max}(P)$, then:

$$\frac{d}{dt} \| e \| \leq -\frac{\varepsilon}{2\gamma} \| e \|$$

$$\| e(t) \| \leq \sqrt{\frac{\gamma}{\alpha}} \| e(0) \| \, exp \left(-\frac{\varepsilon}{2\gamma} t \right)$$

By making $\xi = \sqrt{\frac{\gamma}{\alpha}} \| e(0) \|$ and $\lambda = \frac{\varepsilon}{2\gamma} t$:

$$\| e(t) \| \leq \xi exp \left(-\lambda t \right)$$

$\square$

This result shows that the transmitters oscillator states can be reconstructed and the reconstruction error will decrease as time evolves.

5.3.3   Liouvillian System Properties Based Receiver

The design of this receiver is based on the property that makes a system be Liouvillian (given in Definition 5.1), this property allows to reconstruct the states of a system without the need of a state observer, so it is possible to create receivers that will not be affected by the limitations of observers, and yet they will retain most of their security features, the receiver dynamic is given by:

$$\hat{x}_1 = \int \left(y - f\left[\int (y - d)dt \right] \right)$$

$$\hat{x}_2 = y \qquad\qquad\qquad (5.16)$$

$$\hat{x}_3 = \int (y - d)dt$$

With this receiver the error in the reconstruction of the key stream is zero:

$$\mathbf{e} = \begin{bmatrix} x_1 - \hat{x}_1 \\ x_2 - \hat{x}_2 \\ x_3 - \hat{x}_3 \end{bmatrix} = 0 \qquad\qquad (5.17)$$

$$= \begin{bmatrix} \int \left(y - f\left[\int (y-d)dt \right] \right) dt - \int \left(y - f\left[\int (y-d)dt \right] \right) dt \\ y - y \\ \int (y-d)dt - \int (y-d)dt \end{bmatrix}$$

This reconstruction requires that the initial conditions of the output state x_2 be known, this makes the encryption algorithm slightly different, the initial condition of the output is formed by the first section of the key:

$$x_2(0) = \left(\frac{ABCD}{FFFF} \right) E$$

The remaining initial conditions x_1 and x_3 will be formed by the plain image the same way it was done in the case of the observer, with this modification the trajectories of the chaotic oscillator will depend on the image and key, in consequence if the message changes the order of the shuffling and diffusion values will change as well. An important advantage over the state observer is that there is no error in the reconstruction of the states nor there is a waiting time for the states of the observer to reach the states of the chaotic oscillator.

Observers have an error in the reconstruction of the states, if the error is big enough it could carry over the decryption process, to overcome this the Julia set values are computed with a fixed precision given by a key element ε that contains information about the number of decimals used for the fractal, then it is possible to say that if

$$\|e\| \ll \varepsilon$$

The synchronization error will not affect the remaining part of the algorithm, then from the synchronization error boundary provided by the observers proof of stability the following condition is obtained:

$$\|e\| < \xi \exp(-\lambda t) < \varepsilon$$

The condition refers that if enough time has passed the exponential boundary will be smaller than the desired value ε and in consequence so will be the error. The error of the observer will directly affect the samples taken from the observer $\hat{k}_{si}$, then if this sample error is smaller than the allowed by the Julia set the observer error will not affect the remaining decryption steps, this leads to the expression:

$$\left| k_{si} - \hat{k}_{si} \right| < \varepsilon$$

Where $\hat{k}_{si}$ is the ith sample retrieved from the observer, these can be expressed as $\hat{k}_{si} = k_{si} + e_i$ which is the ith value generated by the oscillator plus the error given by the observer in that specific sample, so:

$$|k_{si} - k_{si} - e_i| < \varepsilon$$

$$\xi \exp(-\lambda t) < \varepsilon$$

$$\ln\left[\left(\frac{\varepsilon}{\xi} \right)^{-\frac{1}{\lambda}} \right] < t$$

So the exponential observer guarantees that the error will not affect the data decryption after $\frac{1}{\lambda} ln \left(\frac{\varepsilon}{\xi} \right)^{-1}$ seconds have passed and in the case of the reconstruction based receiver waiting is not necessary as the error is zero at all time instants. It is also important to mention that if an observer does not guarantees an error smaller than ε the decrypted data will not be accurate.

5.4   Numerical Simulation

Experimental results of the proposed encryption scheme are shown in this section using an RGB color image. The image encryption is based on a Colpitts oscillator that generates the shuffling and diffusion information. In the stage of decryption, the parameters of the slave (the receiver system) are set to the values given by the key, thanks to the successful chaotic synchronization obtained between master and slave systems the encrypted data can be recovered. For the numerical simulation the exponential polynomial observer gains are computed by means of the inequality $PK_iC > 0$, resulting in $k_1 = [0.16, 16, 0.8]^T$ and $k_2 = [0.91, 2.15, 0]^T$.

The results obtained by both receivers are presented in Figs. 5.2 to 5.6. The encrypted image generated with the observer is given by Fig. 5.3. The encrypted image generated by the Liouvillian reconstruction is given in Fig. 5.4. The observer produces the image shown in Fig. 5.5. The Liouvillian reconstruction provides the recovered image shown in Fig. 5.6.

Fig. 5.2 Plain image

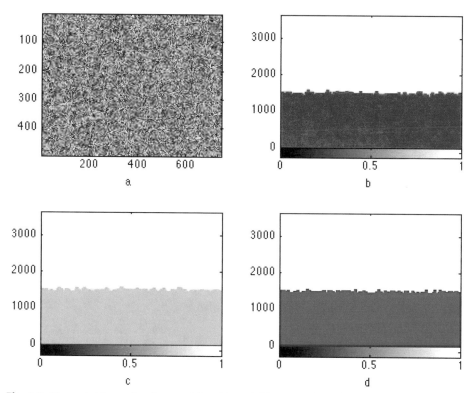

Fig. 5.3 Encrypted image by observer (**a**) and its red (**b**), green (**c**) and blue (**d**) histograms

Figures 5.2 and 5.3 show that the message is properly hidden, this is seen in the histograms as the equal number of pixels with different intensity levels in red, green and blue. The error in the reconstruction of the states provided by the observer, forces us to use less precision while computing the diffusion and shuffling vectors, slightly reducing the ability of the algorithm to withstand brute force attacks, in the case of the Liouvillian reconstruction this is not necessary as there is no error in the reconstruction of the states.

5.5   Concluding Remarks

Using the chaotic nature of the oscillator allowed to produce great variation in the vectors used for shuffling and diffusion depending on the message and by using a state observer and the Liouvillian properties removed the limitation of having to provide more data than the key and cipher, this feature greatly enhances the resistance to chosen plaintext or known plaintext attacks which are one of the biggest threats to this type of algorithm. Using the Julia set to reduce the relation between the key and the cypher image also strengthens the security as the values provided by the norm of a complex number can

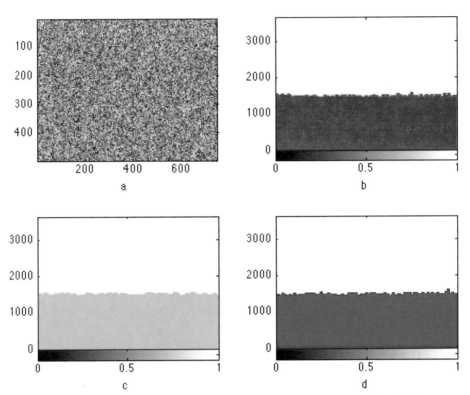

Fig. 5.4 Encrypted image by the reconstruction (**a**) and its red (**b**), green (**c**) and blue (**d**) histograms

be formed by various different sets of real and complex parts, so a very large amount of numbers can generate the same values thus making impossible to determine which one was used further difficulting cryptanalysis. The properties of Liouvillian systems allowed to use more decimals in the reconstruction of the states of the oscillator, in consequence using larger keys is feasible. Another important advantage of the Liouvillian systems is that there is no need to wait for the observer states to converge to the oscillators states, then the implementation is faster, the vectors can be generated immediately, also, the implementation of the transmitter is much more simple thus making the whole encryption and decryption process much faster. It must be remarked that all this benefits are achieved without sacrificing any of the security features of the observer.

Fig. 5.5 Recovered image by the observer (**a**) and its red (**b**), green (**c**) and blue (**d**) histograms

Fig. 5.6 Recovered image by the reconstruction (**a**) and its red (**b**), green (**c**) and blue (**d**) histograms

References

1. Chua, L. O., Kocarev, L., Eckert, K., & Itoh, M. (1992). Experimental chaos synchronization in Chua's circuit. *International Journal of Bifurcation and Chaos*, *2*(3), 705–708.
2. Alvarez, G., & Li, S. (2006). Some basic cryptographic requirements for chaos-based cryptosystems. *International Journal of Bifurcation and Chaos*, *16*(8), 2129–2151 (2006).
3. Yong-Ping, Z., & Shu-Tang, L. (2008). Gradient control and synchronization of Julia sets. *Chinese Physics B*, *17*, 543–549.
4. Hassan, M. F. (2014). Observer design for constrained nonlinear systems with application to secure communication. *Journal of the Franklin Institute*, *351*(2), 1001–1026.
5. Castro-Ramírez, J., Martínez-Guerra, R., & Cruz-Victoria, J. C. (2015). A new reduced-order observer for the synchronization of nonlinear chaotic systems: An application to secure communications. *Chaos: An Interdisciplinary Journal of Nonlinear Science*, *25*(10), 103128.
6. Martínez-Guerra, R., Gómez-Cortés, G. C., & Pérez-Pinacho, C. A. (2015). *Synchronization of integral and fractional order chaotic systems a differential algebraic and differential geometric approach with selected applications in real-time*. Springer.

Fractional Systems

6

Abstract

This chapter focuses on properties, classical results, and applications of Gamma and Beta functions. The theory presented in this chapter is useful for the development of fractional calculus theory and the solution of fractional differential equations, presented in the following chapters.

6.1 Gamma Function

Definition 6.1 Let f be a piecewise continuous function on every finite interval in $[0, \infty)$. If there exist constants $M > 0$ and α such that

$$|f(t)| \leq M e^{\alpha t}, \forall t \in [0, \infty). \tag{6.1}$$

Then it is said that $f(t)$ has **exponential order** α.

Definition 6.2 Let $f(t)$ be an arbitrary function with exponential order; then the integral

$$F(s) = \mathscr{L}\{f(t)\} = \int_0^\infty e^{-st} f(t)dt \tag{6.2}$$

is called the **Laplace transform** of $f(t)$.

Remark 6.1 The exponential order α of a function $f(t)$ guarantees the existence of $\mathscr{L}\{f(t)\}$ for $s > \alpha$, and the Lerch's Theorem can justify the uniqueness.

© The Author(s), under exclusive license to Springer Nature Switzerland AG 2023
R. Martínez-Guerra et al., *Encryption and Decryption Algorithms for Plain Text and Images using Fractional Calculus*, Synthesis Lectures on Engineering, Science, and Technology, https://doi.org/10.1007/978-3-031-20698-6_6

Exercise 6.1 Show that $f(t) = t^n$ has exponential order α for any $\alpha > 0$, $n \in \mathbb{N}$.

According to the previous exercise, consider the integral

$$I = \int_0^\infty e^{-st} t^{n-1} dt.$$

It is not difficult to show, with $\tau = st$ that

$$I = \frac{1}{s^n} \int_0^\infty e^{-\tau} \tau^{n-1} d\tau = \frac{K(n)}{s^n}$$

where the integral $K(n)$ depends only on n and appears in many theoretical and applied problems in science and engineering.

Definition 6.3 The function $\Gamma : (0, \infty) \to \mathbb{R}$ given by

$$\Gamma(z) = \int_0^\infty t^{z-1} e^{-t} dt \tag{6.3}$$

is called **gamma function**.

Although the function (6.3) is defined for $z \in (0, \infty)$, this domain can be extended to a more general set, for example the complex numbers $\mathbb{C}$. The only condition is that $Re(z) \in \mathbb{R} \setminus (\mathbb{Z}^- \cup \{0\})$ since in the zero and in the negative integers, the Gamma function has simple poles. To show this, note that

$$\Gamma(z) = \underbrace{\int_0^1 t^{z-1} e^{-t} dt}_{(1)} + \underbrace{\int_1^\infty t^{z-1} e^{-t} dt}_{(2)} \tag{6.4}$$

For the integral (1):

$$\int_0^1 t^{z-1} e^{-t} dt = \int_0^1 \sum_{k=0}^\infty \frac{(-t)^k}{k!} t^{z-1} dt = \sum_{k=0}^\infty \frac{(-1)^k}{k!} \int_0^1 t^{k+z-1} dt = \sum_{k=0}^\infty \frac{(-1)^k}{(k+z)k!}$$

This series is convergent except where one of the terms has a simple pole, i.e., at $z = -k$, $k \geq 0$, with residue $(-1)^k/k!$. On the other hand, the integral (2) is an entire function, so that $\Gamma(z)$ is a meromorphic function [1].

Example 6.1 Using the definition of the gamma function, calculate $\Gamma(1)$.

Solution

$$\Gamma(1) = \int_0^\infty t^{1-1} e^{-t} dt = \lim_{k \to \infty} \int_0^k e^{-t} dt = \lim_{k \to \infty} \left[-e^{-t} \right]_0^k = \lim_{k \to \infty} \left[-e^{-k} + 1 \right] = 1$$

6.1.1 Some Properties of the Gamma Function

Now, we present some properties of the Gamma function.

Theorem 6.1 *If* $x > -1$, *then*

$$\Gamma(x + 1) = x\Gamma(x) \tag{6.5}$$

Proof Directly from the definition and using integration by parts:

$$\Gamma(x + 1) = \int_0^\infty t^x e^{-t} dt$$

$$= \lim_{z \to \infty, y \to 0^+} \int_y^z t^x e^{-t} dt$$

$$= \lim_{z \to \infty, y \to 0^+} \left[\left(-t^x e^{-t} \right)_y^z + \int_y^z t^{x-1} e^{-t} dt \right]$$

$$= \lim_{z \to \infty, y \to 0^+} x \int_y^z t^{x-1} e^{-t} dt$$

$$= x \int_0^\infty t^{x-1} e^{-t} dt$$

$$= x\Gamma(x)$$

□

Example 6.2 Calculate $\Gamma(2)$.

Solution
By the Theorem 6.1 and the Example 6.1, we have

$$\Gamma(2) = \Gamma(1 + 1) = 1\Gamma(1) = 1$$

Theorem 6.2 *Let* $n \in \mathbb{N}$, *then*

$$(n-1)! = \Gamma(n), \forall n \tag{6.6}$$

Proof Using induction over n. Let A be the set of positive integers such that satisfy (6.6), i.e.,

$$A = \{n \in \mathbb{N} | (n-1)! = \Gamma(n)\}$$

Firstly, we verify that $1 \in A$. Indeed,

$$(1-1)! = 0! = 1 = \Gamma(1).$$

Let us prove the induction step. Suppose that $n \in A$ (induction hypothesis), i.e., $(n-1)! = \Gamma(n)$ and prove that $n+1 \in A$. By the Theorem 6.1 we have that $\Gamma(n+1) = n\Gamma(n)$, then

$$\Gamma(n+1) = n\Gamma(n) = n\,(n-1)! = n! = ((n+1)-1)!$$

We have shown that $n+1 \in A$. Given $1 \in A$ and for every positive integer n belonging to the set A, the number $n+1 \in A$. Then, by using the mathematical induction principle $A = \mathbb{N}$ and (6.6) is valid for every natural number n. For more details about mathematical induction see [2–4]. □

Theorem 6.3 (The Gauss Error Integral)

$$\Gamma(x) = 2 \int_0^{\infty} z^{2x-1} e^{-z^2} dz \tag{6.7}$$

Proof Let $t = z^2$, then $dt = 2zdz$. According to this change of variable, in the integral (6.3), when $t = 0$, $z = 0$ and $z = \infty$ if $t = \infty$. Then

$$\Gamma(x) = \int_0^{\infty} \left(z^2\right)^{x-1} e^{-z^2} 2zdz$$

$$= 2 \int_0^{\infty} z^{2x-1} e^{-z^2} dz$$

$$= 2 \int_0^{\infty} z^{2x-1} e^{-z^2} dz.$$

□

Theorem 6.4

$$\int_0^{\pi/2} \cos^{2x-1}\theta \sin^{2x-1}\theta d\theta = \frac{\Gamma(x)\Gamma(y)}{2\Gamma(x+y)} \tag{6.8}$$

Proof To prove the expression (6.8), let us consider the following integral:

$$\begin{aligned}
\mathscr{I} &= \int_0^\infty \int_0^\infty e^{-\tau^2-v^2}\tau^{2x-1}v^{2y-1}d\tau dv \\
&= \int_0^\infty \tau^{2x-1}e^{-\tau^2}d\tau \int_0^\infty v^{2y-1}e^{-v^2}dv \qquad = \frac{\Gamma(x)}{2}\frac{\Gamma(y)}{2} \tag{6.9}
\end{aligned}$$

where Eq. (6.7) was considered. On the other hand, we consider the integral $\mathscr{I}$ in polar coordinates. Let $\tau = r\cos\theta$ and $v = r\sin\theta$, then

$$\begin{aligned}
\mathscr{I} &= \int\int_R \exp\left(-r^2\cos^2\theta - r^2\sin^2\theta\right)(r\cos\theta)^{2x-1}(r\sin\theta)^{2y-1}rdrd\theta \\
&= \int_0^\infty \int_0^{\pi/2} e^{-r^2}r^{2x-1}\cos^{2x-1}\theta r^{2y-1}\sin^{2x-1}\theta rdrd\theta \\
&= \int_0^\infty e^{-r^2}r^{2(x+y)-1}dr \int_0^{\pi/2}\cos^{2x-1}\theta\sin^{2x-1}\theta d\theta
\end{aligned}$$

Now, by using Eq. (6.7) one has that

$$\mathscr{I} = \frac{1}{2}\Gamma(x+y)\int_0^{\pi/2}\cos^{2x-1}\theta\sin^{2x-1}\theta d\theta \tag{6.10}$$

Finally from expressions (6.9) and (6.10), Eq. (6.8) is direct. This completes the proof.

□

Corollary 6.1

$$\Gamma\left(\frac{1}{2}\right) = \sqrt{\pi} \tag{6.11}$$

Proof In the expression (6.8), let $x = y = \frac{1}{2}$. Then

$$\int_0^{\pi/2} d\theta = \frac{\Gamma\left(\frac{1}{2}\right)\Gamma\left(\frac{1}{2}\right)}{2\Gamma(1)}$$

$$\frac{\pi}{2} = \frac{1}{2}\left[\Gamma\left(\frac{1}{2}\right)\right]^2$$

$$\pi = \left[\Gamma\left(\frac{1}{2}\right)\right]^2$$

and the result follows directly.   □

Corollary 6.2

$$\int_0^\infty e^{-t^2} dt = \frac{\sqrt{\pi}}{2} \tag{6.12}$$

Proof In Eq. (6.7) with $x = 1/2$:

$$\Gamma\left(\frac{1}{2}\right) = 2\int_0^\infty e^{-t^2} dt$$

The proof concludes applying the Corollary 6.1.   □

6.2   Beta Function

Definition 6.4 The beta function is defined as

$$\beta(x, y) = \int_0^1 t^{x-1}(1-t)^{y-1} dt \tag{6.13}$$

Theorem 6.5

$$\beta(x, y) = \frac{\Gamma(x)\Gamma(y)}{\Gamma(x+y)} \tag{6.14}$$

Proof Let $t = \cos^2\theta$. According to this change of variable, the integral (6.13) can be rewritten as follows:

$$\beta(x, y) = \int_0^{\pi/2}\left(\cos^2\theta\right)^{x-1}\left(\sin^2\theta\right)^{y-1} 2\cos\theta \sin\theta d\theta$$

$$= 2\int_0^{\pi/2} \cos^{2x-1}\theta \sin^{2x-1}\theta d\theta \tag{6.15}$$

Finally from the Theorem 6.4, the result is immediate.   □

Remark 6.2 Another way to prove the previous result is through the Laplace transform.
Consider the following convolution

$$h_{x,y}(t) = t^{x-1} * t^{y-1} = \int_0^t (t-\tau)^{y-1}\,\tau^{x-1}d\tau$$

Note that $h_{x,y}(1) = \beta(x, y)$. Applying the Laplace transform to $h_{x,y}(t)$, one has

$$H_{x,y}(s) = \mathcal{L}\left\{h_{x,y}(t)\right\} = \frac{\Gamma(x)}{s^x}\frac{\Gamma(y)}{s^y} = \frac{\Gamma(x)\Gamma(y)}{s^{x+y}}$$

On the other hand, note that

$$\mathcal{L}^{-1}\left\{\frac{\Gamma(x)\Gamma(y)}{s^{x+y}}\right\} = \Gamma(x)\Gamma(y)\mathcal{L}^{-1}\left\{\frac{1}{s^{x+y}}\right\} = \frac{\Gamma(x)\Gamma(y)}{\Gamma(x+y)}t^{x+y-1} = h_{x,y}(t)$$

Therefore, if $t = 1$:

$$\beta(x, y) = \frac{\Gamma(x)\Gamma(y)}{\Gamma(x+y)}$$

According to the axioms of the real numbers, the following result is immediate.

Theorem 6.6

$$\beta(x, y) = \beta(y, x) \tag{6.16}$$

Proof By the equality on the Theorem 6.5 and considering that $\Gamma : (0, \infty) \to \mathbb{R}$, we have

$$\beta(x, y) = \frac{\Gamma(x)\Gamma(y)}{\Gamma(x+y)} = \frac{\Gamma(y)\Gamma(x)}{\Gamma(y+x)} = \beta(y, x)$$

□

Theorem 6.7

$$\beta(x+1, y) = \frac{x}{x+y}\beta(x, y) \tag{6.17}$$

$$\beta(x, y+1) = \frac{y}{x+y}\beta(x, y) \tag{6.18}$$

Proof For (6.17), directly from Eq. (6.14) and using the Theorem 6.1:

$$\beta(x+1, y) = \frac{\Gamma(x+1)\Gamma(y)}{\Gamma(x+y+1)} = \frac{x\Gamma(x)\Gamma(y)}{(x+y)\Gamma(x+y)} = \frac{x}{x+y}\beta(x, y)$$

The proof of (6.18) is similar and it is left to the reader as an exercise.   □

Theorem 6.8 (Legendre Duplication Formula)

$$\Gamma(2x) = \frac{2^{2x-1}}{\sqrt{\pi}}\Gamma(x)\Gamma\left(x + \frac{1}{2}\right) \tag{6.19}$$

Proof Using the definition of Beta function and its equivalence (6.14):

$$\beta(x, x) = \frac{\Gamma(x)\Gamma(x)}{\Gamma(2x)} = \int_0^1 t^{x-1}(1-t)^{x-1} dt$$

Let $t = \frac{1}{2}(1+s)$, then

$$\beta(x, x) = \frac{1}{2}\int_{-1}^1 \frac{1}{2^{x-1}}(1-s)^{x-1}\frac{1}{2^{x-1}}(1+s)^{x-1} ds$$

$$= \frac{1}{2^{2x-1}}\int_{-1}^1 \left(1-s^2\right)^{x-1} ds$$

$$= \frac{2}{2^{2x-1}}\int_0^1 \left(1-s^2\right)^{x-1} ds$$

On the other hand, let $\omega = s^2$, then

$$\beta(x, x) = 2^{2-2x}\int_0^1 (1-\omega)^{x-1}\frac{1}{2}\omega^{-1/2}d\omega$$

$$= 2^{-2x+1}\int_0^1 (1-\omega)^{x-1}\frac{1}{2}\omega^{-1/2+1-1}d\omega$$

$$= 2^{-2x+1}\beta\left(x, \frac{1}{2}\right)$$

Therefore

$$\frac{\Gamma(x)\Gamma(x)}{\Gamma(2x)} = 2^{-2x+1}\frac{\Gamma(x)\Gamma\left(\frac{1}{2}\right)}{\Gamma\left(x+\frac{1}{2}\right)}$$

Considering that $\Gamma\left(\frac{1}{2}\right) = \sqrt{\pi}$, the proof is complete.   □

Corollary 6.3 *Let* $x \in \mathbb{Z}^+$. *Then*

$$\Gamma\left(x + \frac{1}{2}\right) = \frac{(2x)!}{2^{2x}x!}\sqrt{\pi}$$ (6.20)

Proof From the duplication formula (6.19), and applying the property (6.5):

$$\Gamma\left(x + \frac{1}{2}\right) = \frac{\Gamma(2x)\sqrt{\pi}2^{-2x+1}}{\Gamma(x)}$$

$$= \frac{2x\,\Gamma(2x)\sqrt{\pi}}{2^{2x}x\,\Gamma(x)}$$

$$= \frac{\Gamma(2x+1)\sqrt{\pi}}{2^{2x}\,\Gamma(x+1)}$$

Note that $2x + 1 \in \mathbb{Z}^+$ because $x \in \mathbb{Z}^+$. Then applying the relation (6.6), the result is obtained and the proof is complete.   □

An interesting property that establish a relationship with the gamma function and the number π is given in the following result.

Theorem 6.9

$$\Gamma(x)\Gamma(1 - x) = \frac{\pi}{\sin(\pi x)}$$ (6.21)

6.3   Euler's Number and Its Relation to the Gamma Function

According to the Euler's number

$$e^{-t} = \lim_{n \to \infty}\left(1 - \frac{t}{n}\right)^n$$

the Gamma function (6.3) is written as follows:

$$\Gamma(x) = \lim_{n \to \infty}\int_0^n \left(1 - \frac{t}{n}\right)^n t^{x-1}dt$$

Let $t = ns$, then

$$\Gamma(x) = \lim_{n \to \infty} n^x \underbrace{\int_0^1 (1-s)^n s^{x-1} ds}_{\beta(n+1,x)}$$

On the other hand,

$$\beta(n+1, x) = \frac{\Gamma(n+1)\Gamma(x)}{\Gamma(n+x+1)} = \frac{n!(x-1)!}{(x+n)!} = \frac{n!}{x(x+1)(x+2)\dots(x+n)}$$

so that,

$$\Gamma(x) = \lim_{n \to \infty} \frac{n^x n!}{x(x+1)(x+2)\dots(x+n)}$$

$$= \frac{1}{x} \lim_{n \to \infty} \left(\frac{1}{x+1}\right)\left(\frac{2}{x+2}\right)\left(\frac{3}{x+3}\right)\dots\left(\frac{n}{x+n}\right) \cdot \frac{2^x \cdot 3^x \dots (n-1)^x n^x}{1^x \cdot 2^x \dots (n-2)^x (n-1)^x}$$

Note that

$$\frac{2^x \cdot 3^x \dots (n-1)^x n^x}{1^x \cdot 2^x \dots (n-2)^x (n-1)^x} = \left(\frac{2}{1}\right)^x \left(\frac{3}{2}\right)^x \left(\frac{4}{3}\right)^x \dots \left(\frac{n-1}{n-2}\right)^x \left(\frac{n}{n-1}\right)^x$$

$$= \left(1+\frac{1}{1}\right)^x \left(1+\frac{1}{2}\right)^x \left(1+\frac{1}{3}\right)^x \dots \left(1+\frac{1}{n-2}\right)^x \left(1+\frac{1}{n-1}\right)^x$$

and

$$\frac{n}{x+n} = \left(\frac{x+n}{n}\right)^{-1} = \left(1+\frac{x}{n}\right)^{-1}$$

Therefore, the gamma function can be rewritten as follows

$$\Gamma(x) = \frac{1}{x} \prod_{n=1}^{\infty} \left[\left(1+\frac{1}{n}\right)^x \left(1+\frac{x}{n}\right)^{-1}\right] \qquad (6.22)$$

Based in the previous definition of the Gamma function, we proportionate the following alternative proof for Theorem 6.1.

$$
\frac{\Gamma(z+1)}{\Gamma(z)} = \frac{\frac{1}{z+1}\prod_{n=1}^{\infty}\left[\left(1+\frac{1}{n}\right)^{z+1}\left(1+\frac{z+1}{n}\right)^{-1}\right]}{\frac{1}{z}\prod_{n=1}^{\infty}\left[\left(1+\frac{1}{n}\right)^{z}\left(1+\frac{z}{n}\right)^{-1}\right]}
$$

$$
= \frac{z}{z+1}\frac{\lim_{N\to\infty}\prod_{n=1}^{N}\left(1+\frac{1}{n}\right)\lim_{N\to\infty}\prod_{n=1}^{N}(n+z)}{\lim_{N\to\infty}\prod_{n=1}^{N}(n+z+1)}
$$

where

$$
\lim_{N\to\infty}\prod_{n=1}^{N}\left(1+\frac{1}{n}\right) = \lim_{N\to\infty}\prod_{n=1}^{N}\left(\frac{n+1}{n}\right)
$$

$$
= \lim_{N\to\infty}\frac{\prod_{n=1}^{N}(n+1)}{\prod_{n=1}^{N}n}
$$

$$
= \lim_{N\to\infty}\frac{2\cdot3\cdot4\cdot5\ldots(N+1)}{1\cdot2\cdot3\cdot4\ldots N}
$$

$$
= \lim_{N\to\infty}\frac{N!(N+1)}{N!}
$$

$$
= \lim_{N\to\infty}(N+1)
$$

and

$$
\frac{\lim_{N\to\infty}\prod_{n=1}^{N}(n+z)}{\lim_{N\to\infty}\prod_{n=1}^{N}(n+z+1)} = \lim_{N\to\infty}\frac{(1+z)(2+z)(3+z)\ldots(N+z)}{(2+z)(3+z)\ldots(N+z)(N+z+1)}
$$

$$
= \lim_{N\to\infty}\frac{1+z}{N+z+1}
$$

Therefore

$$
\frac{\Gamma(z+1)}{\Gamma(z)} = \frac{z}{z+1}\lim_{N\to\infty}\frac{(N+1)(z+1)}{N+z+1}
$$

$$
= z\lim_{N\to\infty}\frac{N+1}{N+z+1}
$$

$$
= z
$$

since $\lim_{N\to\infty}\frac{N+1}{N+z+1} = 1$, so that $\Gamma(z+1) = z\Gamma(z)$ as in Theorem 6.1.

Lemma 6.1 *Let $b > a$, $m > 0$, $n > 0$. Then*

$$\int_a^b (b - x)^{m-1} (x - a)^{n-1}\, dx = (b - a)^{m+n-1}\, \beta(m, n). \tag{6.23}$$

Proof Let $x = a + y(b - a)$. Then $(b - x)^{m-1}(x - a)^{n-1} = (b - a)^{m+n-2}(1 - y)^{m-1} y^{n-1}$, $dx = (b - a)dy$, and

$$\begin{cases} y = 0, & \text{if} \quad x = a \\ y = 1, & \text{if} \quad x = b \end{cases}$$

Therefore

$$\int_a^b (b - x)^{m-1} (x - a)^{n-1}\, dx = (b - a)^{m+n-1} \int_0^1 (1 - y)^{m-1}\, y^{n-1} dy = \beta(m, n).$$

$\square$

Lemma 6.2 *Let $m > -1$, $p > -1$, $n > 0$. Then*

$$\int_0^1 x^m \left(1 - x^n\right)^p dx = \frac{1}{n} \beta \left(\frac{m + 1}{n}, p + 1\right). \tag{6.24}$$

Proof Let $y = x^n$, then $dy = nx^{n-1}dx$. Hence

$$\int_0^1 x^m \left(1 - x^n\right)^p dx = \int_0^1 x^m (1 - y)^p \frac{dy}{nx^{n-1}}$$

$$= \frac{1}{n} \int_0^1 x^{m-n+1} (1 - y)^p\, dy$$

$$= \frac{1}{n} \int_0^1 y^{\frac{m+1}{n} - 1} (1 - y)^p\, dy$$

$$= \frac{1}{n} \beta \left(\frac{m + 1}{n}, p + 1\right)$$

$\square$

6.4   Miscellaneous Examples

To show the application of the previous theory about gamma and beta functions, we now provide a list of solved exercises.

Exercise 6.2 Prove that $\int_0^\infty e^{-ax} x^n dx = \frac{1}{a^{n+1}} \Gamma(n+1)$, for $n > -1, a > 0$.

Solution

Let $ax = u$, then

$$\int_0^\infty e^{-ax} x^n dx = \int_0^\infty e^{-u} \left(\frac{u}{a}\right)^n \frac{du}{a}$$

$$= \frac{1}{a^{n+1}} \int_0^\infty e^{-u} u^n du$$

$$= \frac{1}{a^{n+1}} \Gamma(n+1)$$

□

Exercise 6.3 Prove that $\int_0^\infty x^m e^{-x^n} dx = \frac{1}{n} \Gamma\left(\frac{m+1}{n}\right)$, for $m > -1, n > 0$.

Solution

Let $u = x^n$, then

$$\int_0^\infty x^m e^{-x^n} dx = \int_0^\infty u^{m/n} e^{-u} \frac{du}{n} u^{(1-n)/n}$$

$$= \frac{1}{n} \int_0^\infty e^{-u} u^{\frac{m+1}{n}-1} du$$

$$= \frac{1}{n} \Gamma\left(\frac{m+1}{n}\right)$$

□

Exercise 6.4 Prove that $\int_a^\infty e^{2ax-x^2} dx = \frac{\sqrt{\pi}}{2} e^{a^2}$.

Solution

Let $u = x - a$, then $u^2 = (x-a)^2 = x^2 - 2ax + a^2$. Based on this change of variable, if $x = a$, then $u = 0$, and for $x = \infty$, $u = \infty$. Besides $2u du = 2(x-a)dx$ and

$$dx = \frac{u du}{x-a} = \frac{x-a}{x-a} du = du$$

Therefore, and using the Corollary 6.2:

$$\int_a^\infty e^{2ax-x^2} dx = \int_0^\infty e^{a^2-u^2} du$$

$$= e^{a^2} \int_0^\infty e^{-u^2} du$$

$$= e^{a^2} \frac{\sqrt{\pi}}{2}$$

□

Exercise 6.5   Prove that $\int_0^{\pi/2} \tan^n \theta \, d\theta = \frac{1}{2} \Gamma\left(\frac{1+n}{2}\right) \Gamma\left(\frac{1-n}{2}\right)$, for $|n| < 1$.

Solution

It is clear that $\tan^n \theta = \cos^{-n} \theta \sin^n \theta$. Then, by using the Theorem 6.4 with $x = \frac{1-n}{2}$ and $y = \frac{1+n}{2}$:

$$\int_0^{\pi/2} \tan^n \theta \, d\theta = \int_0^{\pi/2} \cos^{-n} \theta \sin^n \theta \, d\theta$$

$$= \int_0^{\pi/2} \cos^{2\left(\frac{1-n}{2}\right)-1} \theta \sin^{2\left(\frac{n+1}{2}\right)-1} \theta \, d\theta$$

$$= \frac{1}{2} \frac{\Gamma\left(\frac{1+n}{2}\right) \Gamma\left(\frac{1-n}{2}\right)}{\Gamma(1)}$$

The result is immediate considering that $\Gamma(1) = 1$.

□

Exercise 6.6   Prove that $\int_0^{\pi/2} \sin^n \theta \, d\theta = \int_0^{\pi/2} \cos^n \theta \, d\theta = \frac{\sqrt{\pi}}{2} \frac{\Gamma\left(\frac{n+1}{2}\right)}{\Gamma\left(\frac{n+2}{2}\right)}$.

Solution

In Eq. (6.8), if $x = \frac{1}{2}$ and $y = \frac{n+1}{2}$, then

$$\int_0^{\pi/2} \sin^n \theta \, d\theta = \frac{\Gamma\left(\frac{1}{2}\right) \Gamma\left(\frac{n+1}{2}\right)}{2\Gamma\left(\frac{n+2}{2}\right)}$$

On the other hand, if $x = \frac{n+1}{2}$ and $y = \frac{1}{2}$, then

$$\int_0^{\pi/2} \cos^n \theta d\theta = \frac{\Gamma\left(\frac{1}{2}\right) \Gamma\left(\frac{n+1}{2}\right)}{2\Gamma\left(\frac{n+2}{2}\right)}$$

Finally considering that $\Gamma\left(\frac{1}{2}\right) = \sqrt{\pi}$, the result is immediate. □

Exercise 6.7 Solve $\int_{-\infty}^{\infty} 3x^3 \left(x^3 + 1\right)^2 e^{-x^6 - 2x^3} dx$.

Solution
From the Theorems 6.1 and 6.3, for $x = \frac{3}{2}$ we have

$$\Gamma\left(\frac{3}{2}\right) = 2 \int_0^{\infty} z^2 e^{-z^2} dz = \Gamma\left(\frac{1}{2} + 1\right) = \frac{1}{2}\Gamma\left(\frac{1}{2}\right) = \frac{\sqrt{\pi}}{2}$$

On the other hand, let $u^2 = \left(x^3 + 1\right)^2 = x^6 + 2x^3 + 1$, then

$$\int_{-\infty}^{\infty} 3x^2 \left(x^3 + 1\right)^2 e^{-x^6 - 2x^3} dx = e \int_{-\infty}^{\infty} e^{-u^2} u^2 du$$

$$= e \left[2 \int_0^{\infty} e^{-u^2} u^2 du \right]$$

$$= \frac{e\sqrt{\pi}}{2}$$

Exercise 6.8 Solve the integral $\mathscr{I} = \int_0^{\pi/2} \sqrt{\tan \theta} d\theta$.

Solution
Firstly, note that

$$\sqrt{\tan \theta} = \sqrt{\frac{\sin \theta}{\cos \theta}} = (\sin \theta)^{1/2} (\cos \theta)^{-1/2}$$

then by using the Theorem 6.4 with $x = \frac{1}{4}$ and $y = \frac{3}{4}$:

$$\mathscr{I} = \frac{1}{2} \frac{\Gamma\left(\frac{1}{4}\right) \Gamma\left(\frac{3}{4}\right)}{\Gamma(1)} = \frac{1}{2}\Gamma\left(\frac{1}{4}\right) \Gamma\left(1 - \frac{1}{4}\right)$$

Finally, by using the Theorem 6.9 we have

$$\mathscr{I} = \frac{1}{2} \frac{\pi}{\sin\left(\frac{\pi}{4}\right)} = \frac{\sqrt{2}}{2} \pi$$

Exercise 6.9 Solve the integral $\mathscr{I} = \int_0^1 \frac{dx}{\sqrt[3]{1-x^3}}$.

Solution
The integral can be written as follows

$$\mathscr{I} = \int_0^1 \left(1 - x^3\right)^{-1/3} dx$$

If we make the substitution $t = x^3$, we obtain $dt = 3x^2 dx$ and $x = \sqrt[3]{x}$. Therefore, by using the beta function and Eq. (6.21):

$$\mathscr{I} = \frac{1}{3} \int_0^1 (1-t)^{-1/3} t^{-2/3}$$

$$= \frac{1}{3} \beta\left(\frac{2}{3}, \frac{1}{3}\right)$$

$$= \frac{1}{3} \frac{\Gamma\left(\frac{2}{3}\right) \Gamma\left(\frac{1}{3}\right)}{\Gamma(1)}$$

$$= \frac{1}{3} \frac{\Gamma\left(\frac{1}{3}\right) \Gamma\left(1 - \frac{1}{3}\right)}{\Gamma(1)}$$

$$= \frac{1}{3} \frac{\pi}{\sin\left(\frac{\pi}{3}\right)}$$

$$= \frac{2\pi}{3\sqrt{3}}$$

Exercise 6.10 Evaluate $\int_0^\infty t^{-3/2} \left(1 - e^{-t}\right) dt$.

Solution

By using integration by parts and the Gauss error integral:

$$\int_0^\infty t^{-3/2} \left(1 - e^{-t}\right) dt = -2t^{-1/2} \left(1 - e^{-t}\right) \Big|_0^\infty + 2\int_0^\infty t^{-1/2} e^{-t} dt$$

$$= 2\int_0^\infty t^{1/2-1} e^{-t} dt$$

$$= 2\Gamma\left(\frac{1}{2}\right)$$

$$= 2\sqrt{\pi}$$

Note that the first term of the right-hand side in the integral by part is zero.

Exercise 6.11 Calculate $\int_{-1}^1 \left(\frac{1+x}{1-x}\right)^{1/2} dx$.

Solution

The substitution $t = \frac{1}{2}(1 + x)$ is useful because $dx = 2dt$. Further

$$\begin{cases} t = 0, & \text{if } x = -1 \\ t = 1, & \text{if } x = 1 \end{cases}$$

Therefore

$$\int_{-1}^1 \left(\frac{1+x}{1-x}\right)^{1/2} dx = \int_0^1 \left(\frac{2t}{2(1-t)}\right)^{1/2} 2dt$$

$$= 2\int_0^1 (1-t)^{-1/2} t^{1/2} dt$$

$$= 2\beta\left(\frac{1}{2}, \frac{3}{2}\right)$$

$$= 2\frac{\Gamma\left(\frac{1}{2}\right)\Gamma\left(\frac{3}{2}\right)}{\Gamma(2)}$$

By the Theorem 6.1, $\Gamma\left(\frac{3}{2}\right) = \Gamma\left(\frac{1}{2} + 1\right) = \frac{1}{2}\Gamma\left(\frac{1}{2}\right)$, and recalling Example 6.2 we have that

$$\int_{-1}^1 \left(\frac{1+x}{1-x}\right)^{1/2} dx = \left[\Gamma\left(\frac{1}{2}\right)\right]^2 = \pi$$

Exercise 6.12 Prove that $\beta(n, n+1) = \frac{1}{2}\frac{[\Gamma(n)]^2}{\Gamma(2n)}$, and by using this equality, show that

$$\int_0^{\pi/2} \left(\frac{1}{\sin^3\theta} - \frac{1}{\sin^2\theta}\right)^{1/4} \cos\theta \, dx = \frac{\left[\Gamma\left(\frac{1}{4}\right)\right]^2}{2\sqrt{\pi}}.$$

Solution

Using the equality from the Theorem 6.5,

$$\beta(n, n+1) = \frac{\Gamma(n)\Gamma(n+1)}{\Gamma(2n+1)} = \frac{n\Gamma(n)\Gamma(n)}{2n\Gamma(2n)} = \frac{1}{2}\frac{[\Gamma(n)]^2}{\Gamma(2n)}.$$

If $n = \frac{1}{4}$, then

$$\beta\left(\frac{1}{4}, \frac{5}{4}\right) = \frac{1}{2}\frac{\left[\Gamma\left(\frac{1}{4}\right)\right]^2}{\Gamma\left(\frac{1}{2}\right)} = \frac{\left[\Gamma\left(\frac{1}{4}\right)\right]^2}{2\sqrt{\pi}}.$$

On the other hand, from the definition of the beta function

$$\mathscr{I} = \beta\left(\frac{1}{4}, \frac{5}{4}\right) = \int_0^1 (1-t)^{1/4} t^{-3/4} dt.$$

With the substitution $t = \sin\theta$, we obtain $dt = \cos\theta \, d\theta$. Then

$$\mathscr{I} = \int_0^{\pi/2} (\sin\theta)^{-3/4}(1-\sin\theta)^{1/4}\cos\theta \, d\theta = \int_0^{\pi/2}\left(\frac{1-\sin\theta}{\sin^3\theta}\right)^{1/4}\cos\theta \, d\theta$$

This concludes the proof.   □

Exercise 6.13 Simplify the integral $\int_0^1 \left(\frac{1}{x} - 1\right)^{1/4} dx$.

Solution

Employing the properties for the gamma and beta function as in the previous examples, we have

$$\int_0^1 \left(\frac{1}{x} - 1\right)^{1/4} dx = \int_0^1 x^{-1/4}(1-x)^{1/4} dx = \beta\left(\frac{3}{4}, \frac{5}{4}\right)$$

$$= \frac{\Gamma\left(\frac{3}{4}\right)\Gamma\left(\frac{5}{4}\right)}{\Gamma(2)}$$

$$= \Gamma\left(\frac{3}{4}\right)\Gamma\left(1+\frac{1}{4}\right)$$

$$= \frac{1}{4}\Gamma\left(\frac{1}{4}\right)\Gamma\left(\frac{3}{4}\right)$$

$$= \frac{1}{4}\Gamma\left(\frac{1}{4}\right)\Gamma\left(1-\frac{1}{4}\right)$$

$$= \frac{1}{4}\frac{\pi}{\sin\left(\frac{\pi}{4}\right)}$$

$$= \frac{\sqrt{2\pi}}{4}$$

Exercise 6.14 Simplify the integral $\int_0^1 \left(\ln\frac{1}{x}\right)^{a-1} dx, a > 0$.

Solution

The substitution $x = e^{-t}$ is useful because $\frac{1}{x} = e^t$ and $t = \ln\left(\frac{1}{x}\right)$, so that

$$\left(\ln\frac{1}{x}\right)^{a-1} = t^{a-1}$$

and $dx = -e^{-t}dt$. In addition, from the change of variable

$$\begin{cases} t \to \infty, & \text{if } x \to 0 \\ t = 0, & \text{if } x = 1 \end{cases}$$

Therefore

$$\int_0^1 \left(\ln\frac{1}{x}\right)^{a-1} dx = \int_0^\infty t^{a-1}e^{-t}dt = \Gamma(a)$$

6.5 Fractional-Order Differential Equations

In the following, some linear ordinary differential equations of fractional-order will be solve by means of the Laplace transform method.

6.5.1   Laplace Transform of Fractional-Order Functions

Let us remember the formula to obtain the Laplace transform of a function $f(t)$,

$$\mathscr{L} = \int_0^\infty f(t) e^{-st} dt$$

which is used to convert the function from the time domain (t) to the frequency domain (s) [5].

Example 6.3 The Laplace transform of $f(t) = e^{nt}$ is

$$\mathscr{L}\left[e^{nt}\right] = \int_0^\infty e^{nt} e^{-st} dt$$

$$= \int_0^\infty e^{(n-s)t} dt$$

$$= \int_0^\infty e^{-(s-n)t} dt$$

$$= \left[-\frac{1}{s-n} e^{-(s-n)t}\right]_0^\infty$$

$$= -\frac{1}{s-n}\left[e^{-\infty} - e^0\right]$$

$$= -\frac{1}{s-n}[0-1]$$

$$= \frac{1}{s-n}$$

Example 6.4 The Laplace transform of $f(t) = E_{\alpha,\beta}(nt^\alpha)$ is

$$\mathscr{L}\left[E_{\alpha,\beta}\left(nt^\alpha\right)\right] = \int_0^\infty E_{\alpha,\beta}\left(nt^\alpha\right) e^{-st} dt$$

$$= \int_0^\infty \sum_{k=0}^\infty \frac{(nt^\alpha)^k}{\Gamma(\alpha k + \beta)} e^{-st} dt$$

$$= \sum_{k=0}^\infty \frac{n^k}{\Gamma(\alpha k + \beta)} \int_0^\infty t^{\alpha k} e^{-st} dt$$

$$= t^{\beta-1} \sum_{k=0}^{\infty} \frac{n^k}{\Gamma(\alpha k + \beta)} \int_0^{\infty} t^{\alpha k} e^{-st} dt$$

$$= \sum_{k=0}^{\infty} \frac{n^k}{\Gamma(\alpha k + \beta)} \int_0^{\infty} t^{\alpha k + \beta - 1} e^{-st} dt$$

Consider the following change of variable

$$x = st$$

$$dx = sdt$$

Then we have

$$\mathscr{L}\left[E_{\alpha,\beta}\left(nt^\alpha\right)\right] = \sum_{k=0}^{\infty} \frac{n^k}{\Gamma(\alpha k + \beta)} \int_0^{\infty} \left(\frac{x}{s}\right)^{\alpha k + \beta - 1} e^{-x} \frac{1}{s} dx$$

$$= \sum_{k=0}^{\infty} \frac{n^k}{\Gamma(\alpha k + \beta)} \frac{1}{s^{\alpha k + \beta}} \int_0^{\infty} x^{\alpha k + \beta - 1} e^{-x} dx$$

$$= \sum_{k=0}^{\infty} \frac{n^k}{\Gamma(\alpha k + \beta)} \frac{1}{s^{\alpha k + \beta}} \Gamma(\alpha k + \beta)$$

$$= \sum_{k=0}^{\infty} n^k \frac{1}{s^{\alpha k + \beta}}$$

$$= s^{-\beta} \sum_{k=0}^{\infty} (ns^{-\alpha})^k$$

$$= \frac{s^{-\beta}}{1 - ns^{-\alpha}}$$

$$= \frac{s^{\alpha - \beta}}{s^\alpha - n}$$

In the following, some relevant functions and integral operators and their corresponding Laplace transforms are shown.

Observe the similarities between the Laplace transform of the integer and the fractional-order functions, remembering that the Mittag-Leffler function is a generalization of the exponential function. Moreover, the Laplace transform of the differential and integral operators is shown, as well as the Laplace transform of the convolution between two time functions.

Table 6.1 Laplace transform
of some basic functions

$f(t)$	$F(s)$	$f(t)$	$F(s)$
t^n	$\frac{n!}{s^{n+1}}$	$E_\alpha(-at^\alpha)$	$\frac{s^{\alpha-1}}{s^\alpha+a}$
t^α	$\frac{\Gamma(\alpha+1)}{s^{\alpha+1}}$	$t^\alpha E_{1,1+\alpha}(-at)$	$\frac{s^{-\alpha}}{s+a}$
e^{-at}	$\frac{1}{s+a}$	$t^{\alpha-1}E_{\alpha,\alpha}(-at^\alpha)$	$\frac{1}{s^\alpha+a}$
$t^n e^{-at}$	$\frac{n!}{(s+a)^{n+1}}$	$t^{\beta-1}E_{\alpha,\beta}(-at^\alpha)$	$\frac{s^{\alpha-\beta}}{s^\alpha+a}$

Table 6.2 Laplace transforms of integral operators

$f(t)$	$F(s)$
$f(t) * g(t) = \int_0^t f(\tau)g(t-\tau)d\tau = g(t)*f(t) = \int_0^t g(\tau)f(t-\tau)d\tau$	$F(s)G(s)$
$I^n f(t)$	$\frac{F(s)}{s^n} = s^{-n}F(s)$
$^{RL}I^\alpha f(t)$	$s^{-\alpha}F(s)$

6.5.2 Solution of FODE by Means of the Laplace Transform

In this section, a couple of linear fractional-order ordinary differential equations (FODE) will be solve by means of the Laplace transform, with the aid of the formulas given in Tables 6.1 and 6.2.

Example 6.5 Consider the following linear homogeneous FODE

$$^{RL}_0D_t^{1/2}f(t) + af(t) = 0, \quad {}_0D_t^{-1/2}f(0) = C$$

Observe the fractional-order initial conditions required for this problem. Applying the Laplace transform, we have

$$s^{1/2}F(s) - f^{-1/2}(0) + aF(s) = 0$$

$$s^{1/2}F(s) - C + aF(s) = 0$$

$$\left(s^{1/2} + a\right)F(s) = C$$

$$F(s) = \frac{C}{s^{1/2} + a}$$

Finally, applying inverse Laplace transform, we have

$$f(t) = Ct^{-1/2}E_{\frac{1}{2},\frac{1}{2}}\left(-at^{-1/2}\right)$$

Example 6.6 Consider the following linear nonhomogeneous FODE

$$\,_0^{RL}D_t^Q f(t) + \,_0^{RL}D_t^q f(t) = h(t)$$

Applying the Laplace transform to this equation, and considering $0 < q < Q < 1$, we have

$$s^Q F(s) - f^{Q-1}(0) + s^q F(s) - f^{q-1}(0) = H(s)$$

$$(s^Q + s^q)F(s) = C + H(s)$$

where

$$C = f^{Q-1}(0) + f^{q-1}(0)$$

Hence

$$F(s) = \frac{C + H(s)}{s^Q + s^q}$$

In order to find a suitable Laplace transform for this function, the following arrangement is done

$$F(s) = \frac{C + H(s)}{s^q(s^{Q-q} + 1)}$$

$$= \frac{Cs^{-q}}{s^{Q-q} + 1} + H(s)\frac{s^{-q}}{s^{Q-q} + 1}$$

Consider first the left part. Applying the inverse Laplace transform we have

$$\mathcal{L}^{-1}\left[\frac{Cs^{-q}}{s^{Q-q} + q}\right] = Ct^{Q-1}E_{Q-q,Q}\left(-t^{Q-q}\right)$$

For the right part, note that it comprises a convolution

$$\mathcal{L}^{-1}\left[H(s)\frac{s^{-q}}{s^{Q-q} + 1}\right] = \mathcal{L}^{-1}[H(s)] * \mathcal{L}^{-1}\left[\frac{s^{-q}}{sQ - q + 1}\right]$$

$$= h(t) * t^{Q-1}E_{Q-q,Q}\left(-t^{Q-q}\right)$$

Therefore

$$f(t) = CG(t) + \int_0^t G(t - \tau)h(\tau)d\tau$$

where

$$G(t) = t^{Q-1} E_{Q-q,Q}\left(-t^{Q-q}\right)$$

6.6   Fractional Dynamical System

As it has been mentioned, fractional-order dynamical systems, in contrast with the integer-order ones, have been studied strongly in the last decades. This is due to the great amount of applications and physical phenomena whose dynamics present fractional derivatives and integrals, such as diffusion problems, viscoelasticity, polymeric behaviour, financial systems, biological systems, damped mechanical systems, electric circuits, electrochemistry, rheology, fractals and heat propagation. Particularly, in control theory one of the most important contributions has been the development of generalized PID controllers, as well as other fractional-order controllers such as the CRONE and the fractional-order sliding mode controller. Nowadays almost every kind of controller has been extended to its fractional counterpart.

6.6.1   Commensurate Fractional-Order Systems

There exist different definitions for these kind of systems, e.g. the following, which is found in [6].

Definition 6.5 The fractional differential equation

$$g(x, y(x), D_{*0}^{n_1} y(x), D_{*0}^{n_2} y(x), \ldots, D_{*0}^{n_k} y(x)) = 0$$

with $0 < n_1 < n_2 < \cdots < n_k$ and a certain function g is called commensurate if the numbers $n_1, n_2, \ldots, n_k$ are commensurate, i.e. if the quotients n_μ/n_ν are rational numbers for all $\mu, \nu \in \{1, 2, \ldots, k\}$.

In this case, the author uses this definition because it is related to the traditional use of the concept common in number theory. However, in this book the following definition will be used, from [7].

Definition 6.6 Consider the following model in state space:

$$D^\alpha x = Ax + Bu$$

$$y = Cx$$

where $x \in \mathbb{R}^n$, $u \in \mathbb{R}^r$ and $y \in \mathbb{R}^p$, $\alpha = [\alpha_1, \alpha_2, \ldots, \alpha_n]^T$ is the vector of fractional orders. If $\alpha_1 = \alpha_2 = \cdots = \alpha_n = \alpha \in \mathbb{R}$, the system is called commensurate, otherwise it is an incommensurate system.

For this text purposes, consider the following class of commensurate fractional-order nonlinear systems with unknown inputs:

$$D^\alpha x = g(x, u, f)$$
$$y = h(x, u)$$

where $x \in \mathbb{R}^n$ is the state vector, $u \in \mathbb{R}^m$ is the input (control) vector, $f \in \mathbb{R}^q$ is the unknown input vector, $y \in \mathbb{R}^p$ is the output vector, $\alpha = (\alpha_1, \ldots, \alpha_n)$, g and h are analytic functions. Particularly, in this book $0 < \alpha < 1$ will be used.

6.6.2 Incommensurate Fractional-Order Systems

As stated in the past subsection, if a fractional-order system is not commensurate, it is called an incommensurate system. Consider the following.

Definition 6.7 Consider the following model:

$$D^{\alpha_i} x_i = f_i(x_1, x_2, \ldots, x_n)$$

where $x \in \mathbb{R}^n$, $1 \le i \le n$, $i \in \mathbb{Z}^+$. If $\alpha_i \neq \alpha_j$ for at least one value of i, the system is called incommensurate.

References

1. Kleinz, M., & Osler, T. J. (2000). A child's garden of fractional derivatives. *The College Mathematics Journal, 31*(2), 82–88.
2. Hilfer, R., Butzer, P. L., & Westphal, U. (2010). An introduction to fractional calculus. In *Applications of fractional calculus in physics* (pp. 1–85). World Scientific.
3. Lang, S. (1977). *Complex analysis*. Reading, MA: Addison-Wesley.
4. Martinez-Guerra, R., Martinez-Fuentes, O., & Montesinos-Garcia, J. J. (2019). Algebraic and differential methods for nonlinear control theory. In *Elements of commutative algebra and algebraic geometry*. Springer.
5. Schiff, J. L. (1999). *The Laplace transform. Theory and applications*. New York: Springer.
6. Diethelm, K. (2010). *The analysis of fractional differential equations: An application-oriented exposition using differential operators of caputo type*. Berlin: Springer.
7. Petrás, I. (2011). *Fractional-order nonlinear systems: Modeling, analysis and simulation*. Beijing: Springer.

Fractional-Order Liouvillian Systems and Encryption

7

Abstract

This chapter introduces a method for the numerical estimation of the fractional derivative of a signal, a smoothed sliding modes state observer is used to make the estimation. As application for the estimator a color image encryption algorithm is given, the algorithm is based on the synchronization of fractional chaotic Liouvillian systems and its main characteristics are the capability to keep data safe from the most common types of cryptanalysis and handling large color images while producing no data loss.

7.1 Introduction

Since its introduction by Pecora and Carroll [1], synchronization of chaotic systems has been extensively studied in various fields [2–4], in recent years many studies have emerged from the perspective of fractional calculus [5, 6], using fractional calculus involves new challenges for its differences with traditional calculus, although chaotic system synchronization is one of its many applications, fractional calculus has many others [7–12].

A relevant application to fractional chaotic system synchronization is secure communications, there have been various advances in secure communications that use fractional order chaotic systems [13–23] that are mostly stream ciphers, these ciphers employ many different methods to modify messages, such as logistic maps, DNA coding, transformations and synchronization like coupled systems synchronization, projective synchronization, observer based synchronization and even delayed systems synchronization. There are also block ciphers based on synchronization, although most of them consist on modifications to widely used algorithms such as AES.

© The Author(s), under exclusive license to Springer Nature Switzerland AG 2023
R. Martínez-Guerra et al., *Encryption and Decryption Algorithms for Plain Text and Images using Fractional Calculus*, Synthesis Lectures on Engineering, Science, and Technology, https://doi.org/10.1007/978-3-031-20698-6_7

Some of the mentioned algorithms rely on state observers to achieve synchronization, while this is helpful when dealing with uncertainties, they can have estimation error and if the encrypted data is not handled carefully this error can affect the data reconstruction and cause data loss. Another interesting feature is that not all of the cited ciphers make the numbers used for encryption change if the message is changed, this characteristic can be exploited by an attacker to recover useful data that could finally lead to breaking the algorithm.

There have been many advances in control of chaotic systems [24–28], these proposals could produce good results when applied to secure communications, also interesting publications in sliding modes for fractional systems [29–31] and state estimation of fractional systems [32–43] have been made. Most sliding modes based state observers have the chattering effect, thus making them unsuitable for applications where accuracy on the reconstruction of a state is of most importance, to contribute to these works, the authors propose a smoothed sliding modes state observer for estimation of derivatives, this observer does not produce noticeable chattering and performs very well for the purpose of data reconstruction without error, thanks to the lack of chattering and its resilience to uncertainties in the system dynamic.

Fractional Liouvillian systems are a class of nonlinear fractional chaotic systems, they allow to reconstruct their states without the need of a state estimator, instead the reconstruction is done by using the output, this characteristic is desirable in secure communications, as it reduces the complexity of the receiver as well as the time required to reach synchronization, another remarkable characteristic is that the reconstruction error is zero at all times. There are no reported uses of this type of systems in the literature regarding encryption by fractional chaotic systems, therefore this work presents their usefulness and how to use their more prominent characteristics.

The rest of the chapter is organized as follows: Sect. 7.2 presents preliminaries and important results, Sect. 7.3 contains the fractional derivative estimation via the smoothed sliding modes observer, in Sect. 7.4 the encryption algorithm is given, Sect. 7.5 shows the decryption process, in Sect. 7.6 numerical results are shown, in Sect. 7.7 a security analysis to test the performance of the algorithm is made and finally in Sect. 7.8 some conclusions are made.

7.2   Preliminaries

Before proceeding to the main result and the encryption algorithm, it is necessary to give some important definitions and results, this is to allow a better understanding of the algorithm. The following fractional operators provide the basis for this work:

Definition 7.1 The Riemann-Liouville fractional integral of order $\alpha > 0$, $\alpha \in \mathbb{R}$ is defined as:

$$_0 I_t^\alpha f(t) = \frac{1}{\Gamma(\alpha)} \int_0^t f(\tau)(t - \tau)^{\alpha - 1} d\tau$$

where $\Gamma(\alpha) = \int_0^\infty t^{\alpha - 1} e^{-t} dt$ is Euler's gamma function.

Definition 7.2 The Caputo fractional derivative of order $\alpha > 0$ is defined as:

$$_0^C D_t^\alpha f(t) = \frac{1}{\Gamma(n - \alpha)} \int_0^t f^{(n)}(\tau)(t - \tau)^{n - \alpha - 1} d\tau$$

where $n - 1 < \alpha < n$, $f^{(n)}(\tau)$ is the n-th derivative of $f(t)$.

There is a kind of nonlinear systems that does not fulfill the fractional algebraic observability condition, and yet, it is possible to reconstruct their states:

Definition 7.3 Fractional Liouvillian system: A fractional chaotic system is said to be Liouvillian if its states can be written as a function of Riemann-Liouville fractional integrals of its output, exponentials of Riemann-Liouville fractional integrals of the output and a number of sequential derivatives of its output.

The ability to reconstruct states without relying on state observers nor requiring derivatives can be used to obtain a reconstruction with minimal or no error, this is a desirable feature in secure communications.

Lemma 7.1 ([44]) *A system has the equilibrium point $x = 0$ and $\mathbb{D} \subset \mathbb{R}$ is a domain that contains the origin, let $V(t, x(t)) : [0, \infty) \times \mathbb{D} \to \mathbb{R}$ be a continuously differentiable function and locally Lipschitz on x such that:*

$$\alpha_1 \|x(t)\|^a \leq V(t, x(t)) \leq \alpha_2 \|x(t)\|^{ab}$$
$$_0^C D_t^\beta V(t, x(t)) \leq -a_3 \|x(t)\|^{ab}$$

Where $a, b, \alpha_1, \alpha_2, \alpha_3 > 0$, $t \geq 0$, $x \in \mathbb{D}$ and $0 \leq \beta \leq 1$ are real numbers, then equilibrium $x = 0$ is Mittag-Leffler stable and in consequence the system is asymptotically stable.

Proof Having

$$_0^C D_t^\beta V[t, x(t)] \leq -\frac{\alpha_3}{\alpha_2} V[t, x(t)]$$

There is a non negative function $F(t)$ that satisfies:

$$_0^C D_t^\beta V[t, x(t)] + F(t) = -\frac{\alpha_3}{\alpha_2} V[t, x(t)]$$

The Laplace transform makes

$$s^\beta V(s) - V(0) s^{\beta-1} + M(s) = -\frac{\alpha_3}{\alpha_2} V(s)$$

With the positive constant $V(0) = V[0, x(0)]$ and $V(s) = \mathcal{L}[V(t, x[t])]$ causes

$$V(s) = \frac{V(0) s^{\beta-1} - M(s)}{s^\beta + \frac{\alpha_3}{\alpha_2}}$$

If the initial condition is $x(0) = 0$ then $V(0) = 0$ and the solution of the system $_0^C D_t^\alpha x(t) = f(t, x)$ is $x = 0$, if $x(0) \neq 0$ then $V(0) > 0$ as it is Lipschitz from x, hence the solution is:

$$V(t) = V(0) E_\beta\left(-\frac{\alpha_3}{\alpha_2} t^\beta\right) - M(t)\left[t^{\beta-1} E_{\beta,\beta}\left(-\frac{\alpha_3}{\alpha_2} t^\beta\right)\right]$$

Knowing that $t^{\beta-1}$ and $E_{\beta,\beta}\left(-\frac{\alpha_3}{\alpha_2} t^\beta\right)_{\beta,\beta}$ are non negative:

$$V(t) \leq V(0) E_\beta\left(-\frac{\alpha_3}{\alpha_2} t^\beta\right)$$

Then

$$\|x(t)\| \leq \left[\frac{V(0)}{\alpha_1} E_\beta\left(-\frac{\alpha_3}{\alpha_2} t^\beta\right)\right]^{1/a}$$

Since $\frac{V(0)}{\alpha_1} > 0$ for $x(0) \neq 0$, making $d = \frac{V(0)}{\alpha_1} \geq 0$ where $d = 0$ if $x(0) = 0$. The function $V(t, x)$ is locally Lipschitz with respect to x, then $V[0, x(0)] = 0$ if $x(0) = 0$, also $d = \frac{V(0)}{\alpha_1}$ is Lipschitz with respect to $x(0)$ causing that $d(0) = 0$, then:

$$\|x(t)\| \leq \left[d E_\beta\left(-\frac{\alpha_3}{\alpha_2} t^\beta\right)\right]^{1/a}$$

This implies the Mittag-Leffler stability of the system.

7.3   Fractional Derivative Numerical Estimation

Various applications of fractional chaotic systems need the fractional derivatives of a signal or state, these derivatives should be obtained by methods that does not increase noise, does not add any other detrimental effects to signal quality, be as less computing power demanding as possible and maintain good accuracy.

A state observer can be used to estimate fractional derivatives and fulfill the mentioned requirements, consider that the signal S and its $\alpha - th$ derivative ${}_{0}^{C}D_{t}^{\alpha}S$ can be represented by the two state system:

$$
{}_{0}^{C}D_{t}^{\alpha}x_1 = x_2 = S
$$
$$
{}_{0}^{C}D_{t}^{\alpha}x_2 = {}_{0}^{C}D_{t}^{\alpha}S
$$
$$
y = x_1
$$

So a state observer can estimate both states, and thus provide an estimate of the derivative of the input signal S, in order to use this method, the next assumptions must be satisfied:

Assumption 7.1 *The signal S is bounded by a positive real number $\|S\| \leq S_{max}$, $0 < S_{max} < \infty$.*

Assumption 7.2 *The fractional derivative of order α of signal S is bounded by a positive real number $\|{}_{0}^{C}D_{t}^{\alpha}S\| \leq S'_{max}$, $0 < S'_{max} < \infty$.*

Assumption 7.3 *There exists solutions $P = P^{T} \geq 0$, $Q = Q^{T} > 0$ to the inequality $(A - KC)^{T}P + P(A - KC) < -Q$ where $\lambda_{min}(Q) > \Pi$, also $PB = \rho_1 C^{T}$, $e^{T}PB \leq p_r \|e\|$ and $k_3 e^{T}PB \leq \Lambda \|e\| + \Pi \|e\|^{2}$ with the positive real numbers $0 < \rho_1 < \infty$, $0 < p_r < \infty$. $0 < \Lambda < \infty$ and $0 < \Pi < \infty$.*

Theorem 7.1 *The smoothed sliding modes numerical estimator given by:*

$$
{}_{0}^{C}D_{t}^{\alpha}\hat{x}_1 = \hat{x}_2 + k_1 Ce
$$
$$
{}_{0}^{C}D_{t}^{\alpha}\hat{x}_2 = {}_{0}^{C}D_{t}^{\alpha}\hat{S} = k_2 Ce + k_3 \tanh(Ce)
$$
$$
\hat{y} = \hat{x}_1
$$

Is able to provide an estimate of ${}_{0}^{C}D_{t}^{\alpha}S$ if the Assumptions 7.1, 7.2, and 7.3 are satisfied.

Proof The estimation error is:

$$
e = \begin{bmatrix} x_1 - \hat{x}_1 \\ x_2 - \hat{x}_2 \end{bmatrix} = \begin{bmatrix} e_1 \\ e_2 \end{bmatrix}
$$

The estimation error has α-th derivative:

$$_0^C D_t^\alpha e_1 = x_2 - \hat{x}_2 - k_1 Ce$$

$$_0^C D_t^\alpha e_2 = _0^C D_t^\alpha S - k_2 Ce - k_3 \tanh(Ce)$$

The estimation errors αth derivative is rewritten as $_0^C D_t^\alpha e = Ae - B(_0^C D_t^\alpha S - k_3 \tanh(Ce)) - KCe$ with $A = \begin{bmatrix} 0 & 1 \\ 0 & 0 \end{bmatrix}$, $B = \begin{bmatrix} 0 \\ 1 \end{bmatrix}$, $C = \begin{bmatrix} 1 & 0 \end{bmatrix}$ and $K = \begin{bmatrix} k_1 \\ k_2 \end{bmatrix}$. The Lyapunov function $V = \frac{1}{2} e^T Pe$ has fractional derivative:

$$_0^C D_t^\alpha V = \left(_0^C D_t^\alpha e\right)^T Pe + e^T P \left(_0^C D_t^\alpha e\right) + 2 \sum_{k=0}^{\infty} \frac{\Gamma(1+\alpha) \left(_0^C D_t^k e\right)^T \left(_0^C D_t^{\alpha-k} e\right)}{\Gamma(1+k)\Gamma(1-k+\alpha)}$$

Since the derivative $_0^C D_t^W e$, $W = 1, 2, 3, \ldots$ exist and is continuous and bounded, then there is a real number M such that $\left\|_0^C D_t^W e\right\| \leq M$, for the real non integer number $0 < \alpha < 1$ there is an integer N such that $N - 1 < \alpha < N$, allowing $_0^C D_t^{\alpha-\xi} e$, $\xi = 1, 2, 3, \ldots$ to be divided into two parts: $_0^C D_t^{\alpha-\xi} e$, $\xi = 1, 2, \ldots, N-1$ and $_0^C D_t^{\alpha-\xi} e$, $\xi = N, N+1, N+2, \ldots$, The first part yields $\left\|_0^C D_t^{-\xi} x\right\| = \left\| I^\xi x \right\| \leq \bar{W}_{max} \|x\|$, $0 < \bar{W}_{max} < \infty$ and the second part causes that $\left\|_0^C D_t^{\alpha-\xi} x\right\| \leq \bar{W}_{max} L \|x\|$, then $_0^C D_t^{\alpha-\xi} e$, $\xi = N, N+1, N+2, \ldots$ satisfies $\left\|_0^C D_t^{\alpha-k} x\right\| \leq \bar{W}_{max} \|x\|$ and $\left\|_0^C D_t^{\alpha-k} x\right\| \leq \bar{W} \|x\|$ with $\bar{W} = max\{\bar{W}_{max} L, \bar{W}_{max}\}$. the gamma function has the boundaries $0 < L_{min} < |\Gamma(1-\alpha+w_k)|$, $0 < L_{min} < \infty$ and since $\frac{\Gamma(k)}{\Gamma(k+1)} = \frac{1}{k}$ for $k = 1, 2, 3, \ldots$ the series $\sum_{k=1}^{\infty} \frac{1}{\Gamma(1+k)}$ converges, then there is $H > 0$ such that $0 < \sum_{k=1}^{\infty} \frac{1}{\Gamma(1+k)} < H$, making $B_1 = \frac{\Gamma(1+\alpha) M \bar{W} H}{L_{min}}$ allows the Lyapunov function fractional derivative to be:

$$_0^C D_t^\alpha V \leq \left(_0^C D_t^\alpha e\right)^T Pe + e^T P \left(_0^C D_t^\alpha e\right) + 2B_1 \|e\|$$

$$_0^C D_t^\alpha V \leq \left[Ae - B(_0^C D_t^\alpha S - k_3 \tanh(Ce)) - KCe\right]^T Pe$$

$$+ e^T P \left[Ae - B(_0^C D_t^\alpha S - k_3 \tanh(Ce)) - KCe\right] + 2B_1 \|e\|$$

$$_0^C D_t^\alpha V \leq e^T \left[(A-KC)^T P + P(A-KC)\right] e$$

$$+ 2e^T P \left[B(_0^C D_t^\alpha S - k_3 \tanh(Ce)\right] + 2B_1 \|e\|$$

Solving the inequality $(A - KC)^T P + P(A - KC) < -Q$ by properly selecting the values of k_1 and k_2 so it fulfills Assumption 7.1 and by Assumption 7.2:

$$_0^C D_t^\alpha V \leq -e^T Qe + 2e^T PB(_0^C D_t^\alpha S - k_3 \tanh(Ce)) + 2B_1 \|e\|$$

$$_0^C D_t^\alpha V \leq -e^T Qe + 2e^T PB(S'_{max} - k_3 \tanh(Ce)) + 2B_1 \|e\|$$

Knowing that $-\tanh(x) \leq -sign(x) + 1$:

$$_0^C D_t^\alpha V \leq -e^T Qe + 2e^T PB(S'_{max} - k_3 sign(Ce) + k_3) + 2B_1 \|e\|$$

$$\leq -e^T Qe + 2pr S'_{max} \|e\| - 2k_3 \rho_1 e^T C^T sign(Ce) + 2\Lambda \|e\|$$

$$+ 2\Pi \|e\|^2 + 2B_1 \|e\|$$

Using the Rayleigh-Ritz inequality on the term $-e^T Qe$ and considering that $-\|Ce\| \leq -\rho_2 \|e\|$ for a real number $\rho_2 > 0$ makes:

$$_0^C D_t^\alpha V \leq -\lambda_{min}(Q) \|e\|^2 + 2pr S'_{max} \|e\| - 2k_3 \rho_1 \rho_2 \|e\| + 2\Lambda \|e\|$$

$$+ 2\Pi \|e\|^2 + 2B_1 \|e\|$$

By Assumption 7.3:

$$_0^C D_t^\alpha V \leq 2pr S'_{max} \|e\| - 2k_3 \rho_1 \rho_2 \|e\| + 2\Lambda \|e\| + 2B_1 \|e\|$$

$$\leq 2\left(pr S'_{max} + \Lambda + B_1 - k_3 \rho_1 \rho_2\right) \|e\|$$

Choosing the value of k_3 so it complies with the bound $k_3 \geq \frac{pr S'_{max} + \Lambda + B_1}{\rho_1 \rho_2}$ causes the fractional derivative to be:

$$_0^C D_t^\alpha V \leq 0$$

Then from Lemma 7.1 it is possible to conclude that the estimation error is Mittag-Leffler stable, making possible to accurately estimate the derivative of the signal.

7.4 Encryption Algorithm

The encryption algorithm makes use of the properties of fractional chaotic systems, the intention is to allow the algorithm to withstand most types of cryptanalysis, this is done by making the values used for encryption depend on the key and the message, causing that each different message be encrypted by a different set of values.

First a fractional Liouvillian chaotic oscillator is needed, the Chua-Hartley oscillator is a good example of this type of systems, it is given by:

$$
{}^C_0 D^\alpha_t x_1 = \rho \left(x_2 + \frac{x_1 - 2x_1^3}{7} \right)
$$

$$
{}^C_0 D^\alpha_t x_2 = x_1 - x_2 + x_3
$$

$$
{}^C_0 D^\alpha_t x_3 = -\beta x_2
$$

$$
y = x_2
$$

The states of the oscillator can be expressed as a function of the output and some of its fractional integrals and derivatives. The second state is already in terms of the output, i.e. $x_2 = y$, from the dynamic of the third state it is possible to express x_3 as an integral of the output:

$$
x_3 = -\beta I^\alpha x_2
$$

$$
= -\beta I^\alpha y
$$

The first state is obtained from the dynamic of the second state:

$$
{}^C_0 D^\alpha_t x_2 = x_1 - x_2 + x_3
$$

$$
{}^C_0 D^\alpha_t y = x_1 - y - \beta I^\alpha y
$$

$$
x_1 = \beta I^\alpha y + {}^C_0 D^\alpha_t y + y
$$

This allows to represent the states as:

$$
x_1 = \beta I^\alpha y + y + {}^C_0 D^\alpha_t y
$$

$$
x_2 = y
$$

$$
x_3 = -\beta I^\alpha y
$$

This fractional chaotic system is suited for use with the encryption algorithm described next:

1. The first step is to create a key comprised of decimal numbers, grouped in sections of 5 characters each, a section is to be denoted by key_i, $i \in \mathbb{N}$, an example of the key is:

$$
Key = 12345 - 12345 - 12345 - \ldots - 12345
$$

$$
Key = key_1 - key_2 - key_3 - \ldots - key_n
$$

The key above is only for exemplification purposes, all keys must be different and long enough.

It is possible to encrypt most types of data, but for ease of visualization, from now on it is assumed that the message is a color RGB image of size $m \times n$ pixels, the image is formed by three matrices of size $m \times n$, each containing natural numbers in the range of 0–255, the matrices are denoted by R, G and B.

2. In the second step the initial conditions for the oscillator are made, it is necessary to mention that the initial condition of the output is to be formed by the key, as knowing it is essential to the reconstruction process of the states, suppose that the initial condition of the output state is needed to be in the range [0, 1] then it is computed by:

$$x_2(0) = \frac{key_1}{99999} = \frac{12345}{99999}$$

The initial conditions of the other states are computed using the message, it is done by rearranging the matrices R and G into vectors of size mn denoted Rv and Gv, the initial condition is then computed by using the key and the new vectors in the next equation:

$$x_1(0) = \frac{\sum_{a=1}^{mn} Rv_a}{255mn}, \quad x_3(0) = \frac{\sum_{a=1}^{mn} Gv_a}{255mn}$$

producing real numbers between 0 and 1 that can be easily scaled into any required range of values.

3. The third step is to create the remaining parameters for the oscillator using the key, the equation from step 1 is again used for this step, as example, the parameter α is required to be $\alpha \in [0.9, 1]$:

$$\alpha = 1 - 0.1 \left[\frac{key_2}{99999} \right] = 1 - 0.1 \left[\frac{12345}{99999} \right]$$

The process must be repeated for the remaining oscillator's parameters, while considering the required values to achieve chaotic behavior.

4. The carrier data signal (the signal that contains the message denoted by s) is made in this step, the signal is crucial in maintaining zero data loss, it must be made so the algorithm can operate properly, even in the presence of reconstruction error, the image of size mn is converted into a vector of size $3mn$, then each component of the vector is converted into its binary representation making it a vector s of size $24mn$, this vector contains only 0 and 1 as data, it is preferred by the authors to use amplitudes between 0.5 and 8 for the vector s, the vector must be centered in zero, considering that the vector has amplitude of 8, it is rescaled by $8(s - 0.5)$ producing a squared signal where 4 is interpreted as 1 and -4 is interpreted as 0. A more accurate method for recovering the values of the data carrier signal is only to consider the sign of the binary signal s, so

a positive value is 1 and a negative value is 0, allowing to accurately operate in the presence of noise.

The signal s must be further altered by a bitwise XOR operation (denoted by $\oplus$), the required binary number for the modification is obtained with the systems output and the key:

$$y_d(t) = 10\left(2^{48} - 1\right)\left(\frac{key_4}{99999}\right)\left(\frac{\|y(t)\|}{y_{max}}\right)$$

$$= 10\left(2^{48} - 1\right)\left(\frac{12345}{99999}\right)\left(\frac{\|y(t)\|}{y_{max}}\right)$$

y_{max} is a bound of the output described by $\|y(t)\| \leq y_{max}$, $y_d(t)$ is then turned into its binary representation, where the 24 outer bits form a new 24 bit number, putting first the latter 12 bits and last the first 12 bits, then the 24 bit is rescaled into the range$[0, 255]$ and denoted by y_r, finally the data carrier signal is:

$$g(s, y) = s \oplus y_r$$

To make this step easier to understand, Fig. 7.1 shows the process to create an 8 bit number by using the 8 outer bits of a 12 bit integer:

The image shows how the first four outer bits (colored in red) of the 12 bit integer take the position of the last outer bits in the 8 bit integer, then the last 4 outer bits of the 12 bit integer (colored in blue) take the position of the first four outer integers of the 8 bit integer.

5. Now it is necessary to establish an initial time for the transmission of the message and the period of the signal, the next equation gives the initial time based on a suggested desired initial time t_1

$$t_i = t_1 + 10^{-3}\left(\frac{key_5}{99999}\right) = t_1 + 10^{-3}\left(\frac{12345}{99999}\right)$$

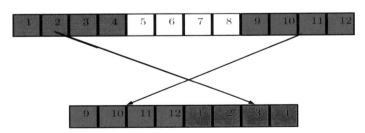

Fig. 7.1 Making an 8 bit integer by using the outer bits

The period of the signal is:

$$t_s = 10^{-3}\left(\frac{key_6}{99999}\right) = 10^{-3}\left(\frac{12345}{99999}\right)$$

Note that the time equations return values in the range [0, 1000] in microseconds, but it can be adjusted as the user requires.

6. Transmit the data carrier signal using the fractional Liouvillian system to mask it, a simple yet efficient method is embedding the signal within one of the states, preferably a state different than the one used for output (choosing a different state has the intention to make cryptanalysis harder), the transmitter for this particular oscillator is:

$$_0^C D_t^\alpha x_1 = \rho\left(x_2 + \frac{x_1 - 2x_1^3}{7}\right) + g\,(s, y)$$

$$_0^C D_t^\alpha x_2 = x_1 - x_2 + x_3$$

$$_0^C D_t^\alpha x_3 = -\beta x_2$$

$$y = x_2$$

Then the output of the oscillator can be used to reconstruct the transmitted data.

7.5 Decryption

The receiver is made using the properties of Liouvillian systems, the structure of such systems makes easy to recover the message s from the output, the equation for this is shown next:

$$\hat{x}_1 = \beta I^\alpha y + y + {}_0^C D_t^\alpha y$$

$$\hat{x}_2 = y$$

$$\hat{x}_3 = -\beta I^\alpha y$$

$$\hat{g}\,(s, y) = {}_0^C D_t^\alpha\left(\beta I^\alpha y + y + {}_0^C D_t^\alpha y\right)$$

$$-\rho\left(y + \frac{\beta I^\alpha y + y + {}_0^C D_t^\alpha y - 2\left(\beta I^\alpha y + y + {}_0^C D_t^\alpha y\right)^3}{7}\right)$$

$$\hat{s} = \hat{g}\,(s, y) \oplus y_r$$

The signal $\hat{g}\,(s, y)$ is obtained using the parts of the key involved in the encryption process. Figure 7.2 shows the encryption and decryption process:

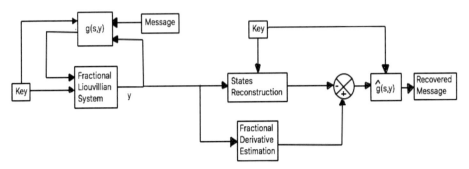

Fig. 7.2 Encryption and decryption

Ideally the properties of Liouvillian systems allow to reconstruct states without error, consider the state reconstruction error:

$$e = \begin{bmatrix} x_1 - \hat{x}_1 \\ x_2 - \hat{x}_2 \\ x_3 - \hat{x}_3 \end{bmatrix} = \begin{bmatrix} e_1 \\ e_2 \\ e_3 \end{bmatrix}$$

The error for each state is:

$$e_1 = x_1 - \hat{x}_1$$
$$= x_1 - \beta I^\alpha y - y - {}^C_0 D^\alpha_t y$$
$$= 0$$
$$e_2 = x_2 - \hat{x}_2$$
$$= x_2 - y$$
$$= 0$$
$$e_3 = x_3 - \hat{x}_3$$
$$= x_3 + \beta I^\alpha y$$
$$= 0$$

The difference between the data carrier signal and the recovered data carrier signal is called the message recovery error, it is denoted by e_s and given by the next expression:

$$e_s = g(s, y) - \hat{g}(s, y)$$
$$= g(s, y) - {}^C_0 D^\alpha_t \left(\beta I^\alpha y + y + {}^C_0 D^\alpha_t y \right)$$

$$-\rho \left(y + \frac{\beta I^\alpha y + y + {}_0^C D_t^\alpha y - 2\left(\beta I^\alpha y + y + {}_0^C D_t^\alpha y\right)^3}{7} \right)$$

$$= 0$$

If the key is known the value y_r can be reconstructed only with the output of the transmitter, then:

$$s - \hat{s} = g\left(s, y\right) \oplus y_r - \hat{g}\left(s, y\right) \oplus y_r = 0$$

The message recovery error is zero, this is subjected to the quality of the reconstruction of the states, it also depends on the ability of the fractional derivative estimation method to produce accurate reconstructions. Once the signal $\hat{s}$ is obtained it is rearranged into its original form, then it yields the reconstructed data and achieves the purpose of zero reconstruction error while providing strong security against cryptanalysis.

7.6 Numerical Results

To test the encryption algorithm a colour image of size 3024×4032 pixels is to be used as message, the image and is RGB histograms can be seen in Fig. 7.3.

The numerical simulation is done using the parameters: $\alpha = 0.92, \rho = 12.75, k_1 = -4$, $k_2 = -5, k_3 = -10$ and $\beta = 100/7$, the initial condition for the output state is $x_2(0) = -0.2$, the other states are $x_1(0) = 0.5$ and $x_3(0) = -0.5$, This values where chosen because they make the oscillator to have a stable dynamic and a double scroll attractor, the convergence of the states and the synchronization error are shown in Fig. 7.4.

The initial error is caused by the small time required for the derivative estimator to reach the value of the derivative, then the slave system synchronizes to the master, the encryption algorithm produces the encrypted image of Fig. 7.5.

The reconstructed image is presented in Fig. 7.6.

The histograms shows that the encryption algorithm is effective in hiding the information, it did not present data loss at all, so it is possible to obtain an accurate reconstruction of the message. The encryption algorithm is versatile enough to encrypt data different than RGB images, a text message is presented next as an example of this:

Message: In this chapter the use of Fractional chaotic Liouvillian systems for secure communications is introduced
Encrypted message: gdagas
Recovered message: In this chapter the use of Fractional chaotic Liouvillian systems for secure communications is introduced

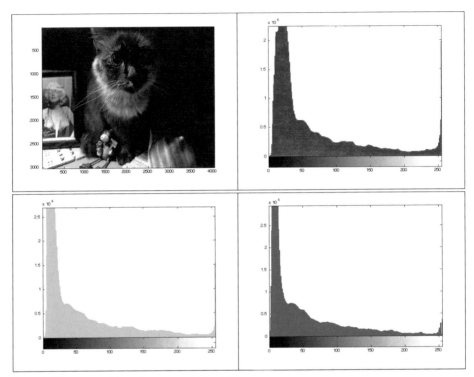

Fig. 7.3 Message and its red, green and blue histograms

The text is accurately reconstructed, although the encryption and decryption process is slightly different than the image the results are the same.

The decryption process relies on the derivative estimate, so it is important that the performance of the chosen method is good, in this case the observer produces an estimate on par with other popular numerical methods, Fig. 7.7 shows the derivative of the output along with the estimates obtained with the observer and Ninteger.

The reconstruction of the derivative is accurate enough to allow the encryption algorithm to correctly recover the plain data without loss, the data carrier signal along the recovered data carrier signal are shown next, for ease of visualization the number 170 is transmitted in its binary representation 10101010:

Figure 7.8 shows that the reconstructed signal does not cause data loss, making an accurate and fast reconstruction of the plain image or plain text.

The performance of the smoothed state observer must be compared to other control methods, first a Luenberger observer given by:

$$ {}_0^C D_t^\alpha \hat{x}_1 = \hat{x}_2 + k_1 C e $$

$$ {}_0^C D_t^\alpha \hat{x}_2 = k_2 C e $$

$$ \hat{y} = \hat{x}_1 $$

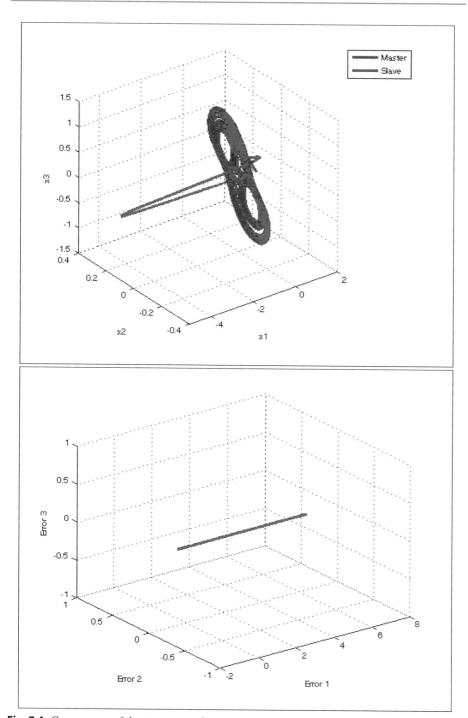

Fig. 7.4 Convergence of the attractors and error

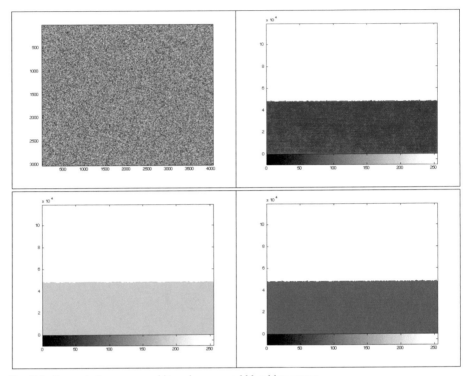

Fig. 7.5 Encrypted message and its red, green and blue histograms

Next a sliding modes state observer described by:

$$_0^C D_t^\alpha \hat{x}_1 = \hat{x}_2 + k_1 sign\,(Ce)$$

$$_0^C D_t^\alpha \hat{x}_2 = k_2 sign\,(Ce)$$

$$\hat{y} = \hat{x}_1$$

Producing the results shown in Fig. 7.9.

The Luenberger observer cannot cope with the presence of the message, it fails to estimate the derivative. The sliding modes observer's chattering causes the reconstruction of the derivative to be inaccurate, and although it resembles the derivative, the reconstruction of the states is useless for decryption, it is possible to say that both control laws are not able to estimate the derivative well enough, making the recovery of the message impossible.

An interesting addition to the analysis of the algorithm is the presence of noise contaminating the transmitted data, the test is done by adding white noise to the output of the function $g\,(s,\,y)$, note that the proof of stability remains the same, as the addition of the noise only increases the value of S'_{max} depending on the amplitude of the noise. The test is performed with two different noise amplitudes, one with around 10% the amplitude

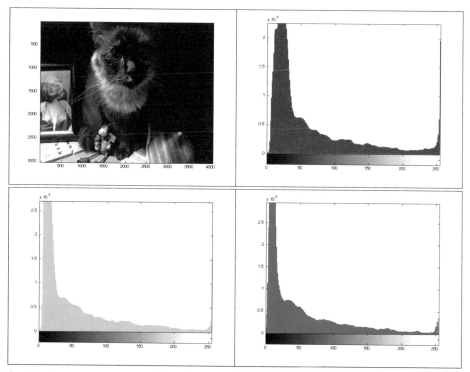

Fig. 7.6 Recovered message and its red, green and blue histograms

of the data carrier signal, and the former with around 25% the amplitude of the data carrier signal, the results can be seen in Figs. 7.10 and 7.11.

To better illustrate the resilience to noise, the 12 megapixel picture is transmitted under the effects of noise contamination, giving the result in Fig. 7.12.

The algorithm can still recover the information without loss, since only the sign of the recovered data carrier signal is employed for the recovery. The encryption algorithm proved capable of managing various types of data while retaining all its features, also the state observer is proven to be able to give good estimates of the fractional derivative of a signal, even in the presence of data contamination by noise.

7.7   Security Analysis

It is necessary to provide an analysis of the algorithm's sensibility to cryptanalysis, for this purpose a known and a chosen plaintext attacks are implemented, also the effectivity of statistical cryptanalysis is tested.

A known plaintext attack is one of the most effective types of cryptanalysis, in it, the attacker knows pairs of message and encrypted message, these are used along with

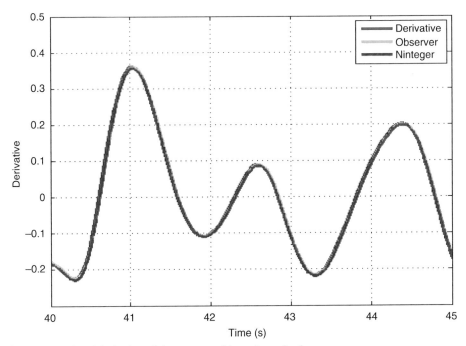

Fig. 7.7 Fractional derivative of the output and its estimated value

knowledge of the algorithm details to recover the key or future messages. In a chosen plaintext attack the attacker has access to the encryption device, the attacker designs messages that produce specific encrypted messages that could lead to breaking the algorithm.

Most of the current encryption algorithms based on fractional chaotic systems do not fare well against these attacks, a typical example of encryption in these works can be found in [10], here the encryption scheme can be expressed by the next equation:

$$ {}_0^C D_t^\alpha x = f(x) $$
$$ y = Cx + s $$

Where the message is s and the encrypted message is y, if the attacker chooses a message such that $s = 0$ (if data is transmitted as amplitudes of the signal, as most commonly is, a black RGB image would cause the desired effect), then the recovered message is $y = Cx$, as the encryption vector Cx depends only on the key, it never changes regardless of the message sent, now if a new message s_2 is sent, recovering without the key is simple: $s_2 = y - Cx = Cx + s_2 - Cx$, now the attacker can easily intercept messages and the key wouldn't even be needed.

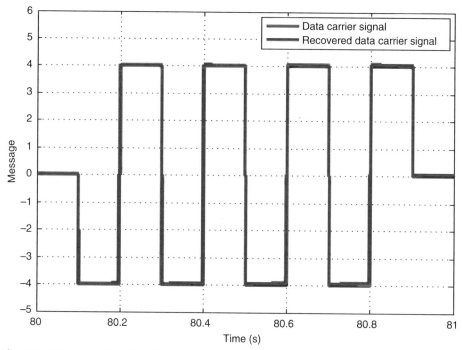

Fig. 7.8 Data carrier signal and the recovered data carrier signal

The proposed encryption algorithm cannot be broken by these attacks, as the encryption vector depends on both, the message and the key, so every different message gets different values for the encryption, rendering the attack useless, even more, the data carrier signal is not directly present in the output of the system, nor in its derivatives, so breaking it is a much more difficult task, to prove this, the mentioned chosen plaintext attack is implemented using a black image of size 3024 × 4032 pixels to try to recover the image from Fig. 7.3, the image used for the attack and the result of it can be seen in Fig. 7.13.

The images above suggest that known and chosen plaintext attacks are ineffective, the attack failed to recover anything resembling the transmitted image in Fig. 7.1, even if it is of the exact same size and implemented with the same initial transmission time and period (for this knowledge of parts of the key is needed), then, the goal of creating an encryption algorithm that can withstand this type of cryptanalysis is achieved.

To further explain the security features, the correlation of two adjacent pixels for the plain image in Fig. 7.3 and its encrypted message from Fig. 7.4 can be found in Fig. 7.14.

The figures show that the plain image correlation of pixels is linear, but the correlation of pixels in the encrypted image is not, the graphics above along the histograms allow to conclude that the plain data is safe from differential and linear cryptanalysis.

For comparison with other encryption algorithms cited in the references, consider the one given in [16], the mentioned method proposes to use a Chen oscillator to mask a data carrier signal, done by embedding it into a state different than the output required

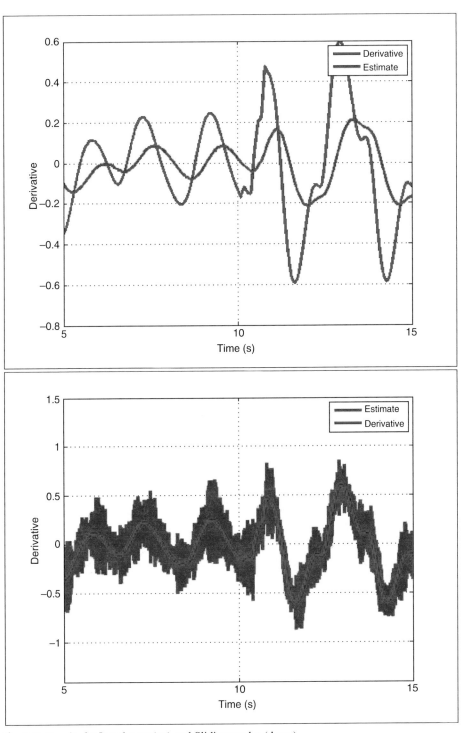

Fig. 7.9 Results for Luenberger(up) and Sliding modes (down)

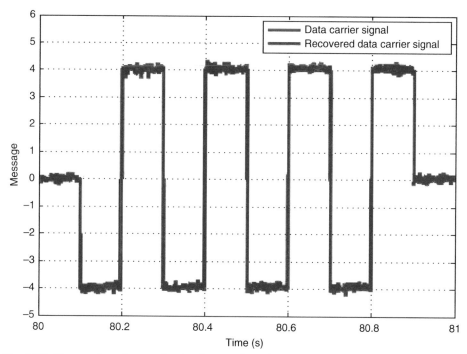

Fig. 7.10 First test for a data carrier signal contaminated by noise

for synchronization, then recover the information by subtracting the proper state from the slave system, this can be better represented by dynamics of the master system:

$$_0^C D_t^\alpha x = f(x)$$

$$y = C_1 x$$

$$y_s = C_2 x + s$$

With $C_1 \neq C_2$. The slave dynamics and message recovery is given by:

$$_0^C D_t^\alpha \hat{x} = f(\hat{x})$$

$$\hat{y} = C_1 \hat{x}$$

$$\hat{s} = y_s - C_2 \hat{x}$$

This encryption scheme is vulnerable to chosen plaintext attacks, the cryptanalysis procedure is the same that was used previously, this is: consider a first message that is designed so the data carrier signal is equal to zero i.e. $s_1 = 0$, the encrypted message is $y_{s1} = C_2 x + s_1 = C_2 x + 0 = C_2 x$, since the values used for the oscillator depend only on

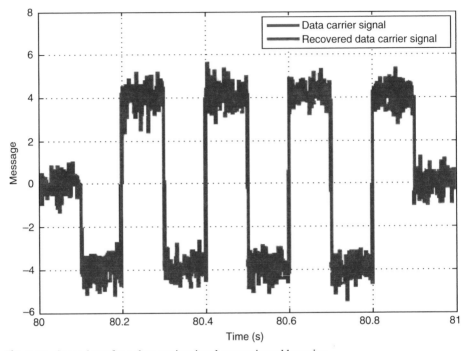

Fig. 7.11 Second test for a data carrier signal contaminated by noise

the key the values of C_2 only change if the key changes, then by recovering the encrypted message y_{s1} allows to recover any subsequent message, this is:

$$y_{sn} = C_2 x + s_n$$

$$s_n = y_{sn} - y_{s1}$$

$$s_n = C_2 x + s_n - C_2 x$$

$$s_n = s_n$$

This shows that it is possible to recover the message without the need of the key. A chosen plaintext attack is implemented on the algorithm of [16], for this purpose, the black image makes the data carrier signal be zero, the results when the second message is a picture can be seen in Fig. 7.15.

The algorithm is unable of maintaining the data safe from the attack, and the second message is recovered without the key. Another setback of this type of encryption is that it uses data carrier signals that consist on the scaled down values of the RGB image pixel values, this method makes the data extremely susceptible to any unknown parameter on the system such as noise, The following result is obtained when the transmitted message is the integer set 71,12, 25, 210, 177, 81, 242, 9 scaled down by 100, noise contamination

Fig. 7.12 Effects of noise on the recovery of an image

to the message is added in the same way that was done in the previous section, producing the results shown in Fig. 7.16.

The noise affects the values recovery, the reconstruction of the set of integers is 65, 22, 5, 234, 155, 110, 255, 0 for the first test and 41, 32, 45, 255, 94, 67, 255, 20 for the second one, showing that it creates an inaccurate reconstruction of the message, causing data loss or a completely wrongful message reconstruction, this is exemplified in Fig. 7.17 by using the same picture as message:

The compared algorithm is unable to retain accuracy in the presence of noise, and even the low amplitude noise used in the previous sections causes severe degeneration of the data. The proposed algorithm is clearly capable to work under such conditions, since a binary data carrier signal is not as affected by noise as a signal formed by scaled down values of the message.

Having done these tests it is possible to say that the encryption algorithm offers good security against most types of cryptanalysis, showing that fractional Liouvillian systems are useful for secure communications.

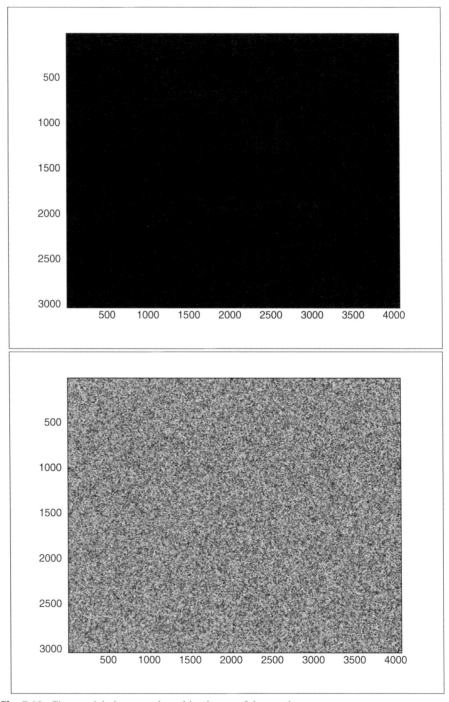

Fig. 7.13 Chosen plain image and resulting image of the attack

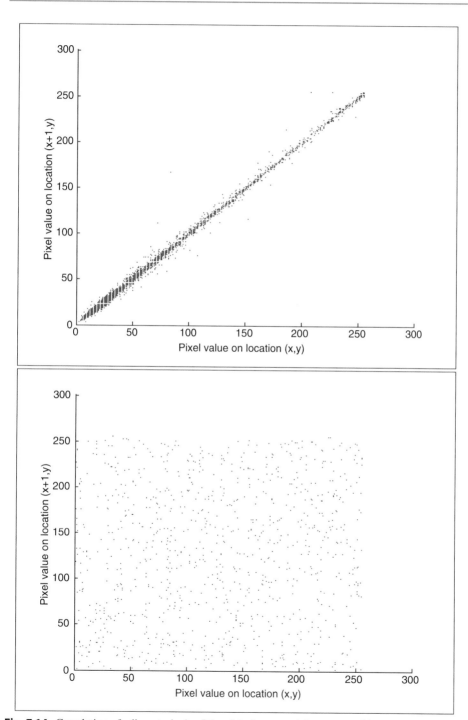

Fig. 7.14 Correlation of adjacent pixels of the plain image and the encrypted image

Fig. 7.15 Chosen plaintext attack result

7.8   Concluding Remarks

Using fractional Liouvillian systems allowed to create an encryption algorithm able to resist cryptanalysis and without data loss, the ability to reconstruct states without error by using only the output is pivotal for achieving this, since the only possible source of error is the implementation of the derivative.

While it is possible to differentiate the required signals directly, doing it causes the noise present on the signal to increase, making the algorithm less reliable, so using another method for estimating the fractional derivative of the signal is necessary, for this task the smoothed sliding modes observer proved accurate and reliable enough, also it did not present the problems of traditional sliding modes derivative approximations when applied to signals with abrupt changes in its trajectory, producing results on par with popular methods for the estimation of the fractional derivative such as Ninteger.

The algorithm proved the effectiveness of fractional chaotic Liouvillian systems when employed in stream ciphers, the decryption process is simple and requires less computing power than most state observers, maintains precision in the reconstruction and gives security to the transmitted messages. The fractional derivative estimator is specifically designed to be as easy to implement and as simple as possible, so it does not affect the performance of the whole stream cipher, finally, the proposed binary signal allowed to

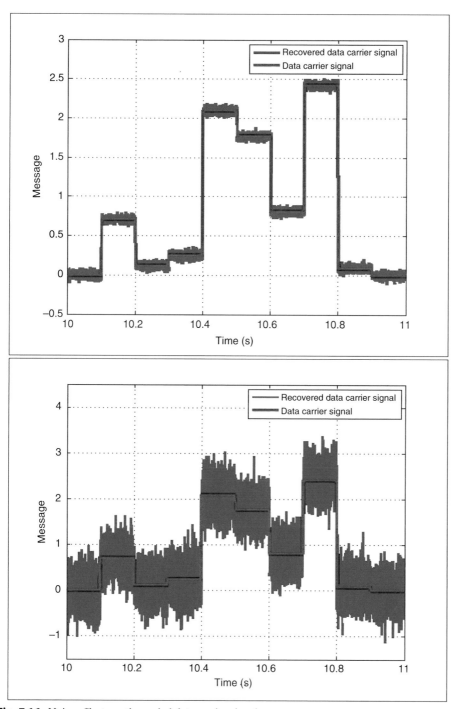

Fig. 7.16 Noise effects on the scaled data carrier signal

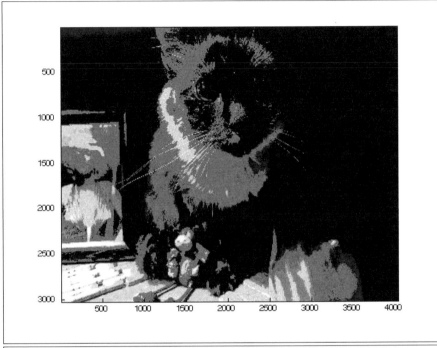

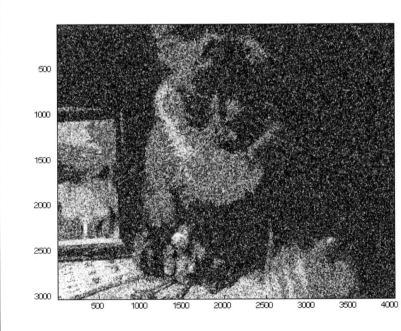

Fig. 7.17 Noise effects on the images

retain the required integrity of the data making possible to combine the data carrier signal with the values from the cryptographic function $g(s, y)$.

References

1. Pecora, L. M., & Carroll, T. L. (1990). Synchronization in chaotic systems. *Physical Review Letters, 64*(8), 821.
2. Pikovsky, A., Rosenblum, M., Kurths, J., & Kurths, J. (2003). *Synchronization: a universal concept in nonlinear sciences* (Vol. 12). Cambridge University Press.
3. Boccaletti, S. (2008). *The synchronized dynamics of complex systems. Monograph series on nonlinear science and complexity* (Vol. 6, pp. 1–239).
4. Balanov, A., Janson, N., Postnov, D., & Sosnovtseva, O. (2008). *Synchronization: from simple to complex.* Springer Science & Business Media.
5. Martínez-Guerra, R., Pérez-Pinacho, C. A., & Gómez-Cortés, G. C. (2015). *Synchronization of integral and fractional order chaotic systems: A differential algebraic and differential geometric approach with selected applications in real-time.* Springer.
6. Li, C., & Chen, G. (2004). Chaos and hyperchaos in the fractional-order Rössler equations. *Physica A: Statistical Mechanics and its Applications, 341*, 55–61.
7. Laskin, N. (2000). Fractional market dynamics. *Physica A: Statistical Mechanics and Its Applications, 287*(3), 482–492.
8. Hamidian, H., & Beheshti, M. T. (2017). A robust fractional-order PID controller design based on active queue management for TCP network. *International Journal of Systems Science, 49*(1), 211–216.
9. Hilfer, R. (Ed.). (2000). *Applications of fractional calculus in physics.* World Scientific.
10. Scalas, E., Gorenflo, R., & Mainardi, F. (2000). Fractional calculus and continuous-time finance. *Physica A: Statistical Mechanics and its Applications, 284*(1), 376–384.
11. Atanackovic, T. M. (2003). On a distributed derivative model of a viscoelastic body. *Comptes Rendus Mecanique, 331*(10), 687–692.
12. Popovic, J. K., Atanackovic, M. T., Pilipovi, A. S., Rapai, M. R., Pilipovi, S., & Atanackovi, T. M. (2010). A new approach to the compartmental analysis in pharmacokinetics: fractional time evolution of diclofenac. *Journal of Pharmacokinetics and Pharmacodynamics, 37*(2), 119–134.
13. N'Doye, I., Darouach, M., & Voos, H. (2013, July). Observer-based approach for fractional-order chaotic synchronization and communication. In *European Control Conference (ECC), 2013* (pp. 4281–4286). IEEE.
14. Luo, C., & Wang, X. (2013). Chaos generated from the fractional-order complex Chen system and its application to digital secure communication. *International Journal of Modern Physics C, 24*(04), 1350025.
15. Wu, X., Wang, H., & Lu, H. (2012). Modified generalized projective synchronization of a new fractional-order hyperchaotic system and its application to secure communication. *Nonlinear Analysis: Real World Applications, 13*(3), 1441–1450.
16. Deng, Y. S., Qin, K. Y., & Shao, S. Q. (2009). Synchronization in coupled fractional order Chen-system and its application in secure communication. In *IEEE International Conference Communications, Circuits and Systems* (pp. 839–841).
17. Kiani-B, A., Fallahi, K., Pariz, N., & Leung, H. (2009). A chaotic secure communication scheme using fractional chaotic systems based on an extended fractional Kalman filter. *Communications in Nonlinear Science and Numerical Simulation, 14*(3), 863–879.

18. Sheu, L. J. (2011). A speech encryption using fractional chaotic systems. *Nonlinear Dynamics*, *65*(1), 103–108.

19. Zhen, W., Xia, H., Ning, L., & Xiao-Na, S. (2012). Image encryption based on a delayed fractional-order chaotic logistic system. *Chinese Physics B*, *21*(5), 050506.

20. Xu, Y., Wang, H., Li, Y., & Pei, B. (2014). Image encryption based on synchronization of fractional chaotic systems. *Communications in Nonlinear Science and Numerical Simulation*, *19*(10), 3735–3744.

21. Kassim, S., Hamiche, H., Djennoune, S., & Bettayeb, M. (2017). A novel secure image transmission scheme based on synchronization of fractional-order discrete-time hyperchaotic systems. *Nonlinear Dynamics*, *88*(4), 2473–2489.

22. Muthukumar, P., Balasubramaniam, P., & Ratnavelu, K. (2015). Fast projective synchronization of fractional order chaotic and reverse chaotic systems with its application to an affine cipher using date of birth (DOB). *Nonlinear Dynamics*, *80*(4), 1883–1897.

23. Zhang, L., Sun, K., Liu, W., & He, S. (2017) A novel color image encryption scheme using fractional-order hyperchaotic system and DNA sequence operations. *Chinese Physics B*, *26*(10) 100504.

24. Xu, Y., Wang, H., Liu, D., & Huang, H. (2015). Sliding mode control of a class of fractional chaotic systems in the presence of parameter perturbations. *Journal of Vibration and Control*, *21*(3), 435–448.

25. Liu, D., Xu, W., & Xu, Y. (2013). Noise-induced chaos in the elastic forced oscillators with real-power damping force. *Nonlinear Dynamics*, *71*(3), 457–467.

26. Xu, Y., Gu, R., Zhang, H., & Li, D. (2012). Chaos in diffusionless Lorenz system with a fractional order and its control. *International Journal of Bifurcation and Chaos*, *22*(04), 1250088.

27. Xu, Y., Gu, R., & Zhang, H. (2011). Effects of random noise in a dynamical model of love. *Chaos, Solitons & Fractals*, *44*(7), 490–497.

28. Xu, Y., Mahmoud, G. M., Xu, W., & Lei, Y. (2005). Suppressing chaos of a complex Duffing's system using a random phase. *Chaos, Solitons & Fractals*, *23*(1), 265–273.

29. Izaguirre-Espinosa, C., Muñoz-Vázquez, A. J., Sánchez-Orta, A., Parra-Vega, V., & Castillo, P. (2016). Attitude control of quadrotors based on fractional sliding modes: theory and experiments. *IET Control Theory & Applications*, *10*(7), 825–832.

30. Aghababa, M. P. (2013). A novel terminal sliding mode controller for a class of non-autonomous fractional-order systems. *Nonlinear Dynamics*, *73*(1–2), 679–688.

31. Kamal, S., Raman, A., & Bandyopadhyay, B. (2013). Finite-time stabilization of fractional order uncertain chain of integrator: An integral sliding mode approach. *IEEE Transactions on Automatic Control*, *58*(6), 1597–1602.

32. Li, C., Wang, J., Lu, J., & Ge, Y. (2014). Observer-based stabilization of a class of fractional order non-linear systems for $0 < \alpha < 2$ case. *IET Control Theory & Applications*, *8*(13), 1238–1246.

33. Zhong, F., Li, H., & Zhong, S. (2016). State estimation based on fractional order sliding mode observer method for a class of uncertain fractional-order nonlinear systems. *Signal Processing*, *127*, 168–184.

34. Aghababa, M. P. (2012). Robust stabilization and synchronization of a class of fractional-order chaotic systems via a novel fractional sliding mode controller. *Communications in Nonlinear Science and Numerical Simulation*, *17*(6), 2670–2681.

35. Boukal, Y., Darouach, M., Zasadzinski, M., & Radhy, N. E. (2017). Robust H_∞ observer-based control of fractional-order systems with gain parametrization. *IEEE Transactions on Automatic Control*, *62*(11), 5710–5723.

36. Lin, C., Chen, B., Shi, P., & Yu, J. P. (2018). Necessary and sufficient conditions of observer-based stabilization for a class of fractional-order descriptor systems. *Systems & Control Letters, 112*, 31–35.
37. Chen, L., Chen, G., Wu, R., Tenreiro Machado, J. A., Lopes, A. M., & Ge, S. (2018). Stabilization of uncertain multi-order fractional systems based on the extended state observer. *Asian Journal of Control, 20*(3), 1263–1273.
38. N'Doye, I., Salama, K. N., & Laleg-Kirati, T. M. (2018). Robust fractional-order proportional-integral observer for synchronization of chaotic fractional-order systems. *IEEE/CAA Journal of Automatica Sinica, 6*(1), 268.
39. Luo, S., Li, S., Tajaddodianfar, F., & Hu, J. (2018). Observer-based adaptive stabilization of the fractional-order chaotic MEMS resonator. *Nonlinear Dynamics, 92*(3), 1079–1089.
40. Yu, W., Li, Y., Wen, G., Yu, X., & Cao, J. (2017). Observer design for tracking consensus in second-order multi-agent systems: Fractional order less than two. *IEEE Transactions on Automatic Control, 62*(2), 894–900.
41. Yang, B., Yu, T., Shu, H., Zhu, D., An, N., Sang, Y., & Jiang, L. (2018). Perturbation observer based fractional-order sliding-mode controller for MPPT of grid-connected PV inverters: Design and real-time implementation. *Control Engineering Practice, 79*, 105–125.
42. Wang, A., Liao, X., & Dong, T. (2018). Fractional-order follower observer design for tracking consensus in second-order leader multi-agent systems: Periodic sampled-based event-triggered control. *Journal of the Franklin Institute, 355*(11), 4618–4628.
43. Coronel-Escamilla, A., Gómez-Aguilar, J. F., Torres, L., Valtierra-Rodriguez, M., & Escobar-Jiménez, R. F. (2017). Design of a state observer to approximate signals by using the concept of fractional variable-order derivative. *Digital Signal Processing, 69*, 127–139.
44. Li, Y., Chen, Y., & Podlubny, I. (2010). Stability of fractional-order nonlinear dynamic systems: Lyapunov direct method and generalized Mittag-Leffler stability. *Computers & Mathematics with Applications, 59*(5), 1810–1821.

Fractional-Order Robust State Observers and Encryption

8

Abstract

This chapter introduces a new encryption algorithm for color RGB images and text. The encryption is based on the synchronization of fractional chaotic systems in a topology of master-slave. The encryption algorithm provides security against common encryption techniques, including known and chosen plain text attacks.

8.1 Introduction

Fractional calculus has gained attention due to its many possible applications in various fields like finance [1], physics [2], medicine [3], biology [4] and chaotic systems synchronization [5], this last one was first introduced by Pecora and Carroll [6] and since then, many possible uses to it were found.

One of the most important applications of systems synchronization is secure communications, it comprises both fractional calculus and synchronization of chaotic systems, there are several publications on image encryption involving chaotic systems and traditional calculus [7–24] and most of them with very good results, but only a few can be found based on fractional calculus [25–32], and in most of these, the safety of the information is not the top priority and fail to provide security against a variety of cryptanalysis, in particular many of the mentioned encryption algorithms will not fare well if chosen plaintext attacks are implemented. In particular references [7] and [32] have interesting encryption algorithms, that use chaotic systems for encryption, in [7] the algorithm uses the key to make the initial conditions and some system parameters, causing that the trajectories from which the encryption values are obtained, vary depending on the key, then, in consequence, the set of numbers used for encryption will also vary depending on

© The Author(s), under exclusive license to Springer Nature Switzerland AG 2023
R. Martínez-Guerra et al., *Encryption and Decryption Algorithms for Plain Text and Images using Fractional Calculus*, Synthesis Lectures on Engineering, Science, and Technology, https://doi.org/10.1007/978-3-031-20698-6_8

the key. Another interesting example that uses fractional chaotic systems can be found on [32], in that work the encryption is done by obtaining pseudorandom numbers from the states of a fractional chaotic oscillator, and with an equation convert these numbers into a fitting range of values to mix them with the plain image. Although the mentioned proposals are good, they fail to provide security against some popular forms of cryptanalysis like chosen plaintext attacks, so the safety of the information encrypted with them is not guaranteed.

The intention of this chapter is to provide an encryption algorithm that uses fractional chaotic systems synchronization while guaranteeing the safety of the information even if chosen plaintext attacks or known plaintext attacks are implemented, to achieve the required security features the properties of chaotic fractional systems will be extensively used and the receiver will be designed as a fractional smoothed sliding modes state estimator.

This chapter is organized as follows: in Sect. 8.2 concepts about fractional calculus are given, in Sect. 8.3 the encryption process and the transmitter are presented, in Sect. 8.4 the observer and the decryption process are explained, in Sect. 8.5 numerical results are shown, in Sect. 8.6 a security analysis of the algorithm is made and finally in Sect. 8.7 conclusions can be found.

8.2   Preliminaries

To further elaborate on the encryption algorithm and the subsequent decryption it is necessary to give the following definitions:

Definition 8.1 The Riemann-Liouville fractional integral of a function $f(t)$ is:

$$_0I_t^\alpha f(t) = \frac{1}{\Gamma(\alpha)} \int_0^t f(\tau)(t-\tau)^{\alpha-1} d\tau$$

With $n-1 < \alpha < n$ and $\Gamma(\bullet)$ is the gamma function given by $\Gamma(\alpha) = \int_0^\infty t^{\alpha-1}e^{-t}dt$ that converges to the right half of the complex plane.

Definition 8.2 The Caputo fractional derivative of order α of a function $f(t)$ is described by:

$$_{t_0}^C D_t^\alpha f(t) = \frac{1}{\Gamma(n-\alpha)} \int_{t_0}^t f^{(n)}(\tau)(t-\tau)^{n-\alpha-1} d\tau$$

And $f^{(n)}(\tau)$ is the n-th derivative, n is positive integer number.

Lemma 8.1 *The Lyapunov function V (x) fulfills the next condition:*

$$a_1 \|x\| \leq V(x) \leq a_2 \|x\|$$

$$D^\alpha V(x) \leq -a_3 \|x\|$$

With $a_1, a_2, a_3 > 0$ the system will be Mittag-Leffler stable and in consequence asymptotically stable.

The encryption algorithm will be based on the synchronization of two fractional chaotic systems, the synchronization will require that the systems be arranged in a master-slave configuration, the transmitter will be the master, which is used to encrypt the data, therefore it will be a fractional chaotic oscillator. The slave will be the receiver, in this case it is designed as a smoothed sliding modes state observer, its function is to synchronize to the master and provide an accurate reconstruction of the states of the master system and with them, reconstruct the message that was encrypted with the master system.

The fractional chaotic oscillator and the state observer will be based on the mentioned operators, also the state observer's proof of stability will use Lemma 8.1. In the following sections a detailed explanation of the encryption and decryption processes are given.

8.3 Encryption Algorithm

The encryption algorithm will encrypt the data carrier signal $g(s, y)$ by combining it with one of the states of the transmitter oscillator:

$$D^\alpha x_1 = x_2$$

$$D^\alpha x_2 = f(x_1, x_2) + g(s, y)$$

$$y = x_1$$

Where $0 < \alpha < 1$, the function $f(x_1, x_2)$ is also bounded by the Lipschitz condition:

$$\|f(a) - f(b)\| \leq L \|a - b\|, \ a, b \in \mathbb{R}, L \in \mathbb{R}^+$$

The output state should be different than the state where $g(s, y)$ was added. For the description of the encryption algorithm it will be assumed that the data is an RGB image of size $i \cdot j$ composed by three matrices containing the red, blue and green data of the image, each matrix is of size $i \cdot j$ and they will be converted into vectors of size (ij) $R \in \mathbb{N}^{ij}$, $0 \leq R_{ij} \leq 255$, $G \in \mathbb{N}^{ij}$, $0 \leq G_{ij} \leq 255$ and $B \in \mathbb{N}^{ij}$, $0 \leq B_{ij} \leq 255$. The data carrier signal $g(s, y)$ is bounded by $\|g(s, y)\| \leq S_{max}$ and $S_{max} > 0$ is a positive real number. The encryption algorithm is the following:

1. Create the encryption key: The key is a set of decimal numbers separated into sections denoted by Key_n, $n = 1, 2, 3, ...$, the key must have enough elements for it to supply the required information for each step of the algorithm, an example of the key is:

$$Key = 12345 - 12345 - 12345 - 12345 - 12345 - ...$$

$$Key = key_1 - key_2 - key_3 - key_4 - key_5 - ...$$

   The keys should be randomly generated and must be long enough for the key space to be large so it deters brute force attacks.

2. The elements of the key must be converted to the values that will make the fractional chaotic oscillator have chaotic behavior, the range of values that will produce chaotic behavior is to be obtained by computing the Lyapunov exponents, a range of values that causes chaos will produce a positive lead Lyapunov exponent, the key elements will be used along the message to create the initial conditions of the oscillator, the key alone will be used for the remaining parameters of the oscillator, the needed values are to be transformed to the proper range as shown next:

$$x_1(0) = \frac{Key_1}{99999} \frac{\sum_{a=1}^{ij} R_a}{255ij}, \quad x_2(0) = \frac{Key_2}{99999} \frac{\sum_{a=1}^{ij} G_a}{255ij}$$

   The utilization of the vectors R and G of the image is only an example, the choice of vectors must be randomized according to a key element, The value 99999 is selected supposing that a range $[0, 1]$ in the state x_1 initial value produces a positive lead Lyapunov exponent, the remaining parameters of the oscillator are computed by:

$$\alpha = \frac{key_3}{99999}$$

   Also considering that the range $\alpha \in [0, 1]$ for the parameter promotes chaos.

3. Create the data carrier signal: In many cases, data is embedded into a the data carrier signal simply by scaling down its values and introducing them into the signal, this is the most easy way to make a data carrier signal, but when working with state observers, and specially sliding modes, in some instances although the error is not noticeable, it will cause that, when scaling up for reconstruction, the small error is carried over and also scaled up, causing data loss and in consequence an erroneous message reconstruction, for example the set of integer values: 170, 55, 32, 255, 8, 45, 123, 15 is scaled down multiplying for 0.01 and recovered by a common sliding modes observer (Fig. 8.1).

   The recovered values are $187, 63, 41, 230, 0, 55, 101, 5$, the inaccurate reconstruction is caused mostly by the chattering inherent to sliding modes, it is necessary to choose a

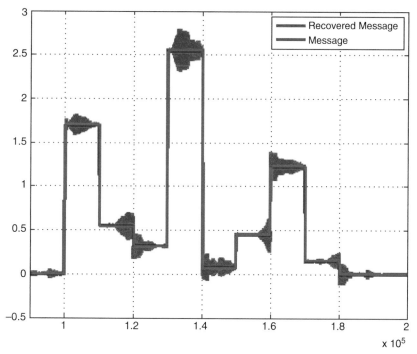

Fig. 8.1 Common sliding modes performance

way of embedding the data that does not cause loss even if there is a slight error in the reconstruction of the signal.

Most data is comprised of 8 bit integer numbers making possible to convert the message into its binary representation, so the information signal will only contain two numbers, 1 and 0. The signal will be a squared waveform of a suggested amplitude of 0.5–1 and centered in zero, this way a 1 will be a positive period and a 0 will be a negative period of the message signal $g(s, y)$, the period P of this signal is given by the encryption key:

$$P = \frac{1}{3 \cdot i \cdot j} \left[\left(t_f + \frac{Key_3}{99999} \right) - \left(t_i + \frac{Key_4}{99999} \right) \right]$$

Where t_i and t_f are the initial and final transmission time respectively. The intention of creating this kind of signal is to avoid data loss, even if there is noise or any other unexpected inaccuracies during the transmission. Figure 8.2 provides an example of an 8 bit integer converted into the signals.

The message must be further masked by combining it with a pseudorandom number obtained from the key and the output, a suggested method is creating a pseudorandom 48 bit number with the systems output by normalizing the output between zero and one,

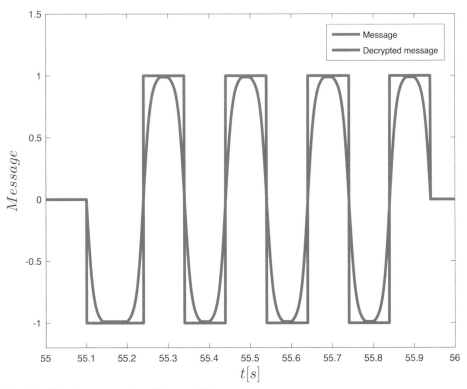

Fig. 8.2 Signal representation of integer 173

multiply by $2^{48} - 1$ and then convert it into a 48 bit binary number which is combined with an element of the key via a bitwise XOR operation. Once the 48 bit binary number is available take the first 12 bits and the last 12 bits of the 48 bit number to create a 24bit number that is again normalized between 0 and 255 and finally turned into a binary 8 bit number y_r, this value is combined with the data carrier signal 8 bit number by a bitwise XOR operation: $g(s, y) = s \oplus y_r$, note that this signal will also be squared with the same characteristics as the messages, this method is suggested by the authors because it yields encrypted images with entropy of around 7.9951 with slight variation caused by the message, key and the systems output, the parts of $g(s, y)$ that do not include information of the message must be filled with random numbers. Finally, the message is transmitted through the output of the system.

The advantage of this encryption scheme is that the message itself is not present in the output nor it requires a separate channel to transmit its data, more over, the signals used for masking the message depend on the key and the message itself, making known and chosen plaintext attacks useless, since each individual combination of key and message will have a very diffident encryption values due to the chaotic nature of the oscillator.

The algorithm can be modified to be used with compressed image files like jpg, in the specific case of jpg there are several possibilities, the most straightforward would be to adjust the range of the pseudorandom values used for encryption to the needs of Y'CrCb that jpg uses and proceed in a similar way with RGB images, another possibility is to directly work with the segments that form the file, this will also require to adjust the values to the proper range, although this could not result as intuitive and easy as the first proposal. It is worth of mention that, if working with compressed files is necessary, the compressed image is to be encrypted, as various types of compressed formats tend to slightly alter the image, then if the RGB image is first encrypted and then compressed, the compression will modify the encrypted message and the decryption process will not be applied to the same data that was encrypted, hence producing an erroneous reconstruction. Also the encryption function is designed to work with RGB images $S \in \mathbb{N}$, and maps positive eight bit integer numbers to positive eight bit integer numbers: $C(S) : \mathbb{N} \to \mathbb{N}$, and so does the decryption algorithm: $D[C(S)] : \mathbb{N} \to \mathbb{N}$, most compression methods do not use only positive integer data, for example the JPEG uses real values: $J_{pg}(S) : \mathbb{N} \to \mathbb{R}$ so if an image is encrypted and then compressed $J_{pg}[C(S)] \in \mathbb{R}$ will produce an erroneous result, because the decryption process is made to work with positive eight bit integers.

8.4   Receiver and Decryption

The receiver will be a smoothed sliding modes state observer, the intention of proposing this kind of observer is to reduce to a minimum the effects of chattering that sliding modes observers have, if chattering is not reduced it could cause data loss if the data signal is not capable of working under such conditions. The receiver dynamics are described by the next equation:

$$D^\alpha \hat{x}_1 = \hat{x}_2 - k_1 \left(y - \hat{y} \right) - k_2 \left| y - \hat{y} \right|^{1/2} \tanh(y - \hat{y})$$
$$D^\alpha \hat{x}_2 = f \left(\hat{x}_1, \hat{x}_2 \right) - k_3 \left(y - \hat{y} \right) - k_4 \tanh \left(y - \hat{y} \right)$$
$$\hat{y} = \hat{x}_1$$
$$\hat{s} = \left(D^\alpha D^\alpha y - f \left(\hat{x}_1, \hat{x}_2 \right) \right) \oplus y_r$$

The synchronization error between the transmitter and the receiver is:

$$e = x - \hat{x}$$
$$e = \begin{bmatrix} x_1 - \hat{x}_1 \\ x_2 - \hat{x}_2 \end{bmatrix}$$

The error's fractional derivative of order α is:

$$D^\alpha e = \begin{bmatrix} e_2 - k_1\,(e) - k_2\,|e|^{1/2}\tanh(e) \\ f\,(x) - f\,(\hat{x}) + g\,(s,y) - k_3\,(e) - k_4\tanh(e) \end{bmatrix}$$

$$D^\alpha e = A + B\phi\,(e) + Bg\,(s,y) - K_a Ce - K_b\tanh(Ce)$$

where $A = \begin{bmatrix} 0 & 1 \\ 0 & 0 \end{bmatrix}$, $B = \begin{bmatrix} 0 \\ 1 \end{bmatrix}$, $\phi\,(e) = f\,(x) - f\,(\hat{x})$ and the gain vectors are $K_a = \begin{bmatrix} k_1 & k_3 \end{bmatrix}^t$ and $k_b = \begin{bmatrix} k_2\,\|e\|^{\frac{1}{2}} & k_4 \end{bmatrix}^T$, the following assumptions will be needed for the proof of stability.

Assumption 8.1 *The next LMI has solutions $P = P^T > 0$ and $Q = Q^T > 0$:*

$$(A - K_a C)^T\,P + P\,(A - K_a C) \le -Q$$

Assumption 8.2 *The inequalities $e^T PK_b \le \frac{1}{2}\lambda_{min}\,(Q)\,\|e\|^2 + \varepsilon\,\|e\|$, $e^T PB \le S_{max}\,\|e\|$ and $e^T PB \le L\,\|e\|$ are true for ε, S_{max}, $L \ge 0$.*

Then, the Lyapunov candidate function and its fractional derivative of order α are:

$$V = e^T Pe$$

$$D^\alpha V = \left(D^\alpha e\right)^T Pe + eP^T\left(D^\alpha e\right)$$

$$+2\sum_{k=0}^{\infty} \frac{\Gamma\,(1+\alpha)\,\left(D^k e\right)^T\left(D^{\alpha-k}e\right)}{\Gamma\,(1+k)\,\Gamma\,(1-k+\alpha)}$$

Since $D^K e$ exist, is continuous and bounded, there is M such that $\left\|D^k e\right\| \le M$ for $k = 1, 2, 3\ldots, \alpha$ is a real non integer number, then, there is an integer N such that $N-1 < \alpha < N$, so $D^{\alpha-k}e$ can be divided into two parts: $D^{\alpha-k}e, k = 1, 2, ..., N-1$ and $D^{\alpha-k}e, k = N, N+1, N+2, ...$, then $D^{\alpha-k}e, k = 1, 2, ..., N-1$ yields $\left\|D^{-k}x\right\| = \left\|I^k x\right\| \le \bar{K}_{max}\,\|x\|$, $\bar{K}_{max} > 0$ and $\left\|D^{\alpha-k}x\right\| \le \bar{K}_{max}L\,\|x\|$, then $D^{\alpha-k}e, k = N, N+1, N+2, ...$ satisfies $\left\|D^{\alpha-k}x\right\| \le \bar{K}_{max}\,\|x\|$, then $\left\|D^{\alpha-k}x\right\| \le K^{-}_{max}\,\|x\|$, with $\bar{K} = max\left\{\bar{K}_{max}L, \bar{K}_{max}\right\}$. The gamma function has the boundaries $0 < L_{min} < |\Gamma\,(1-\alpha+k)|, L_{min} > 0$ and since $\frac{\Gamma(k)}{\Gamma(k+1)} = \frac{1}{k}$ for $k = 1, 2, 3, ...$ the series $\sum_{k=1}^{\infty}\frac{1}{\Gamma(1+k)}$ converges, then there is $H > 0$ such that $0 < \sum_{k=1}^{\infty}\frac{1}{\Gamma(1+k)} < H$, and the Lyapunov functions fractional derivative is:

$$D^\alpha V \le \left(D^\alpha e\right)^T Pe + e^T P\left(D^\alpha e\right) + 2B_1\,\|e\|$$

Where $B_1 = \frac{\Gamma(1+\alpha)M\bar{K}H}{L_{min}}$, then from Assumption 8.2.

$$D^\alpha V \le e^T \left[(A - K_a C)^T P + P (A - K_a C) \right] e$$
$$+ 2e^T P \left[Bg (s, y) + B\phi (e) - K_b \tanh (Ce) \right]$$
$$+ 2B_1 \|e\|$$
$$\le e^T \left[(A - K_a C)^T P + P (A - K_a C) \right] e$$
$$+ 2S_{max} \|e\| + 2L \|e\| - 2e^T P K_b \tanh (Ce)$$
$$+ 2B_1 \|e\|$$
$$\le e^T \left[(A - K_a C)^T P + P (A - K_a C) \right] e$$
$$+ 2 (S_{max} + L + B_1) \|e\| - 2e^T P K_b \tanh (Ce)$$

Using the Assumption 8.1,

$$D^\alpha V \le e^T Q e + 2 (S_{max} + L + B_1) \|e\|$$
$$- 2e^T P K_b \tanh (Ce)$$
$$\le -e^T Q e + 2 (S_{max} + L + B_1) \|e\|$$
$$- 2e^T P K_b \tanh (Ce)$$

Since $-\tanh (x) \le -sign (x) + 1$:

$$D^\alpha V \le -e^T Q e + 2 (S_{max} + L + B_1) \|e\|$$
$$- 2e^T P K_b [sign (Ce) - 1]$$
$$\le -e^T Q e + 2 (S_{max} + L + B_1) \|e\|$$
$$- 2e^T P K_b sign (Ce) + 2e^T P K_b$$
$$\le -e^T Q e + 2 (S_{max} + L + B_1) \|e\|$$
$$- 2e^T P K_b \frac{Ce}{\|Ce\|} + 2e^T P K_b$$
$$\le -e^T Q e + 2 (S_{max} + L + B_1) \|e\|$$
$$- \frac{2}{\sqrt{\lambda_{max} (C^T C)}} e^T P K_b C \frac{e}{\|e\|} + 2e^T P K_b$$

It is known that $\sqrt{\lambda_{max}\left(C^T C\right)} = 1$, λ_{min} and λ_{max} are the minimum and maximum eigenvalues of the argument matrix, then Assumption 8.2 yields:

$$D^\alpha V \leq -e^T Q e + 2\left(S_{max} + L + B_1\right)\|e\|$$
$$-2\lambda_{min}\left(PK_bC\right)\|e\|$$
$$+2\left[\frac{1}{2}\lambda_{min}\left(Q\right)\|e\|^2 + \varepsilon\|e\|\right]$$
$$\leq 2\left[S_{max} + L + B_1 + \varepsilon - \lambda_{min}\left(PK_bC\right)\right]\|e\|$$

By choosing K_b such that $\lambda_{min}\left(PK_bC\right) > S_{max} + L + B_1 + \varepsilon$ makes the Lyapunov functions fractional derivative be:

$$D^\alpha V \leq 0$$

According to lemma 1 the state estimation error is Mittag-Leffer stable, in consequence it will allow to recover the encrypted message by using the key and arranging the binary representation of the data into is original form.

The algorithm for decryption will require the key, it is described in following steps:

1. Synchronize the receiver to the transmitter using the key.
2. Combine the key and the transmitters output to obtain the values of y_r by applying the same operations given in the fourth step of the encryption algorithm.
3. Use $\hat{s} = \left(D^\alpha D^\alpha y - f\left(\hat{x}_1, \hat{x}_2\right)\right) \oplus y_r$ on the recovered binary signal to reconstruct the transmitted data carrier signal.
4. Arrange the recovered binary values in groups of eight numbers, then convert them to 8 bit integer format.

The recovered message accuracy will depend on the observer and the data conversion into a signal, the authors prefer signals of amplitude 0.5–1 for its ease of visualization and comparison with the oscillators states, although the observer is capable of working with far smaller or larger signal amplitudes without data loss.

8.5   Numerical Results

In order to test the performance of the observer and the encryption algorithm, a color image and text will be used as transmitted data. A suited fractional chaotic oscillator would be the fractional Duffing oscillator described by:

$$D^\alpha x_1 = x_2$$

$$D^\alpha x_2 = x_1 - x_1^3 - \delta x_2 + \gamma \cos(\omega t) + g(s, y)$$

$$y = x_1$$

The observer that will act as receiver for this oscillator is:

$$D^\alpha \hat{x}_1 = \hat{x}_2 - k_1 (y - \hat{y})$$

$$-k_2 |y - \hat{y}|^{1/2} \tanh(y - \hat{y})$$

$$D^\alpha \hat{x}_2 = \hat{x}_1 - \hat{x}_1^3 - \delta \hat{x}_2 + \gamma \cos(\omega t)$$

$$-k_3 (y - \hat{y}) - k_4 \tanh(y - \hat{y})$$

$$\hat{y} = \hat{x}_1$$

$$\hat{s} = \left[D^\alpha D^\alpha y \right.$$

$$\left. - \left(\hat{x}_1 - \hat{x}_1^3 - \delta \hat{x}_2 + \gamma \cos(\omega t) \right) \right] \oplus y_r$$

Using the following system parameters $\alpha = 0.973$, $\hat{x}_1(0) = 1$, $\hat{x}_2(0) = -1$, $\delta = 0.6432$, $\omega = 4$ and the gains $k_1 = -8.347$, $k_2 = -2.659$, $k_3 = -3.172$, $k_4 = -4.958$, the next results are obtained when the message is the number 170 (Figs. 8.3, 8.4, and 8.5).

The fractional smoothed sliding modes state observer can accurately recover the message, even if the derivative is involved in the message reconstruction there will not be any data loss, because how the message signal is constructed. To test the image encryption capabilities a 1683×2522 pixels color image is used as message, the message along its RGB histogram can be seen in Fig. 8.6.

The resulting encrypted image along its histograms is presented in Fig. 8.7.

By analyzing the histograms, it is possible to see that the encryption algorithm is effective, the encrypted image histograms show that statistical based cryptanalysis such as linear and differential will not be effective, even more the encryption algorithm renders other types cryptanalysis ineffective, this cryptanalysis techniques being chosen plaintext attacks and known plaintext attacks, a detailed exposition on this security features and resistance to the mentioned attacks can be found on the next section along with the proper tests and a comparison to other state of the art fractional chaotic systems synchronization based algorithms. The decryption process yields the following results (Fig. 8.8).

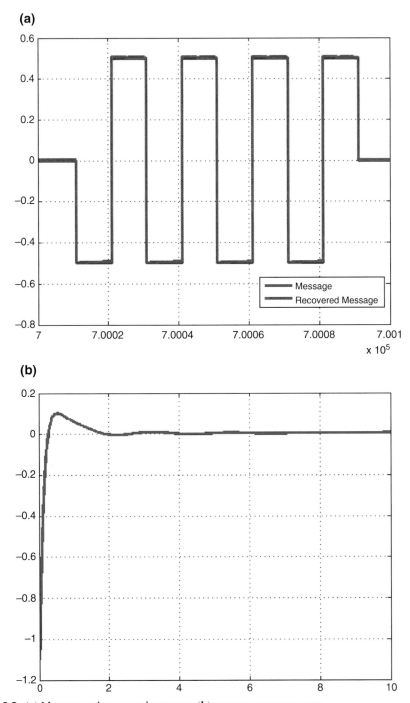

Fig. 8.3 (**a**) Message and recovered message, (**b**) message recovery error

(a)

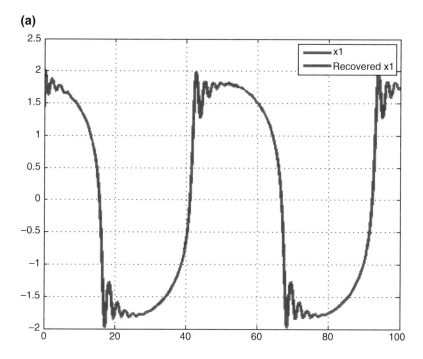

(b)

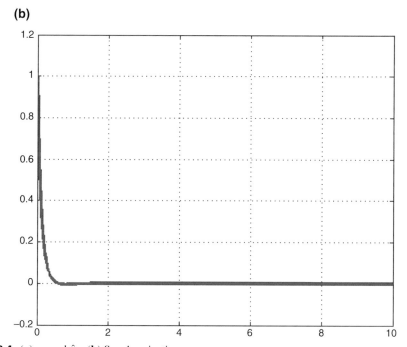

Fig. 8.4 (**a**) x_1 and $\hat{x}_1$, (**b**) Synchronization error

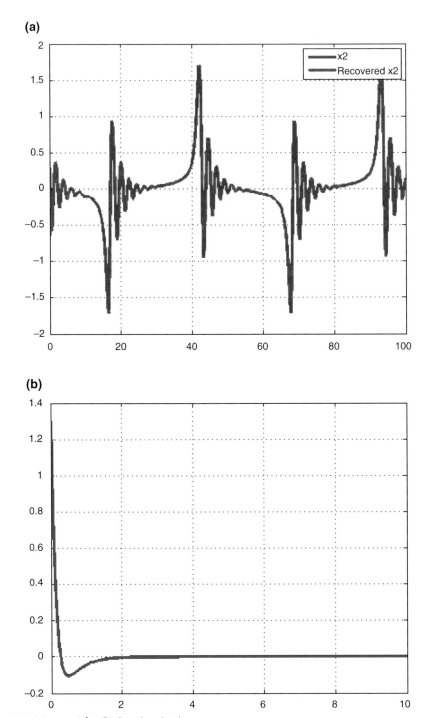

Fig. 8.5 (**a**) x_2 and $\hat{x}_2$, (**b**) Synchronization error

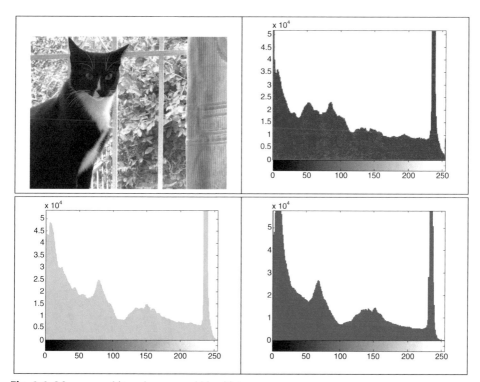

Fig. 8.6 Message and its red, green and blue histograms

The message is recovered without error, the image and its RGB histograms are identical to the transmitted ones, proving that the algorithm can effectively encrypt and decrypt data without any data loss during the process. The algorithm can also encrypt other types of data, such as text messages, in this case, the process is similar and faster than the color image, being different in the creation of initial conditions, to achieve similar results, half of the message creates one initial condition and the remaining half for the other, the obtained results are given next:

Text message: **This chapter introduces a new encryption algorithm that relies on fractional order chaotic systems synchronization**
The encrypted message is:
Encrypted text message: dfasfdasd
And the recovered text message is:
Recovered text message: **This chapter introduces a new encryption algorithm that relies on fractional order chaotic systems synchronization**

The algorithm is capable of encrypting data different than images and maintaining its performance and accuracy in message reconstruction.

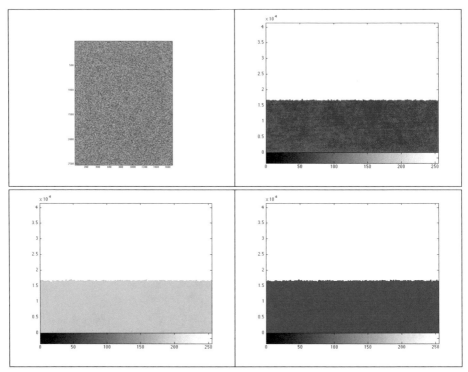

Fig. 8.7 Encrypted message and its red, green and blue histograms

The time computational complexity analysis of the algorithm is obtained according to [33] and [34], the number of pixels composing the image is denoted by ij. The algorithm will require the following number of instructions:

The first step needs one function for providing the key, considering that the range of values for chaotic behavior has been previously found. The second step will use $2ij + N_{pa}$ instructions if the chaotic oscillator has two states, the value of N_{pa} is the number of parameters to be obtained through the key (this number can vary depending on the chaotic oscillator used). The third step will need $24ij + 3$ instructions for creating the binary signal, another $21ij$ the operations done to the signal and finally $24ij$ for transmission, yielding $O(ij) = 71ij + 4 + N_{pa}$. This result shows that the algorithm has linear time complexity. The algorithm was implemented using MATLAB for OS X, with a Core i5 at 1.3Ghz (unable to overclock) and 4GB of DDR3 RAM, the required time varies depending on the type of data, the text message took $1000\,\mu s$, the 1683×2522 RGB image needed $10.302\,s$, the intention of choosing a large color image as message was to show the performance of the algorithm while working with images similar to the ones that are created by current cameras and other image sources, in most publications the authors prefer to use very small greyscale images, even if it is not representative of common image resolutions and sizes, thus giving a non precise representation of the performance of the proposed algorithms.

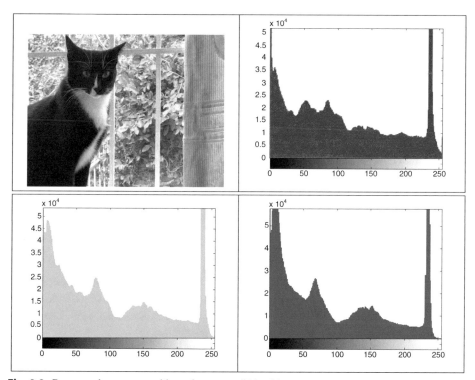

Fig. 8.8 Recovered message and its red, green and blue histograms

In view of the results it is possible to say that the algorithm works well in conjunction with the observer, as it made possible to recover the messages without any data loss, the algorithm proved to be capable of encrypting various types of data of different sizes while providing each of them with the same security features.

8.5.1 Situations that Lead to Decryption Failure

The primary cause of wrong data recovery is to not fulfill the conditions given in the encryption and decryption sections. The binary signal that is proposed is very important, in many works in the literature, there is no specification on how to handle the data carrier signal, the easiest way to create a data carrier signal is to simply scale down the 8 bit integer values and assign to each value a time period, however, when using state observers error will be present, and if such scaled down integer data carrier signal is used, this error though small, will lead to an erroneous reconstruction of the transmitted data, also, if this type of signal is used with the proposed algorithm, the error in the data reconstruction is not guaranteed to be zero. Another mistake that could result in an erroneous reconstruction of the data is to select the gains of the observer different than $\lambda_{min}(PK_bC) > S_{max} + L +$

$B_1 + \varepsilon$, by doing this, the slave system is made unstable, causing the recovered data to be different than the transmitted one. In the next images the results of incurring in such errors can be seen.

Figure 8.9 shows the results of wrongful message reconstruction. In the first image the results of using a different data carrier signal can be seen, in the second, the resulting recovered image of using an unstable state observer is shown. The first case is caused by not using the exact same encrypted values for the bitwise XOR operation causing the decryption to produce a different result, the second erroneous decryption cause is that the states of the observer were not bounded, making impossible to properly recover the data carrier signal.

8.6   Security Analysis

Most of the cited literature regarding encryption by synchronization of fractional chaotic systems can be affected by two cryptanalysis techniques, these two being chosen plaintext attacks and generalized synchronization attacks, to exemplify this, the mentioned attacks are implemented in two algorithms contained in the cited literature, then, these two attacks are applied to our algorithm showing that they are ineffective.

In a known plaintext attack, the attacker knows a previous pair of message and encrypted message, then, the attacker uses this information to extract data from a new encrypted image that could lead to breaking the encryption. The chosen plaintext attack [35] is similar, but with the difference that the attacker can choose what message to send, so the attacker has access to a chosen message and is encrypted message, the pair of messages is used to retrieve information to break the algorithm. In most stream ciphers that use fractional chaotic system synchronization this attack can recover messages without needing the key, the encryption method proposed in [25] is representative of this type of stream ciphers, it can be easily described by the next equations, where the master is:

$$D^\alpha x(t) = Ax(t) + f\left(x(t), \overline{y}(t)\right) + Bd(t) + Ls(t)$$

$$\overline{y}(t) = Cx(t) + s(t)$$

$$x(0) = x(0)$$

and the slave is:

$$D^\alpha \hat{x}(t) = A\hat{x}(t) + f\left(\hat{x}(t), \overline{y}(t)\right) + Bd(t) + L\left[\overline{y}(t) - \hat{y}(t)\right]$$

$$\hat{y}(t) = C\hat{x}(t)$$

$$\hat{x}(0) = x(0)$$

$$\hat{s}(t) = \overline{y}(t) - \hat{y}(t)$$

Fig. 8.9 Improper message
reconstruction

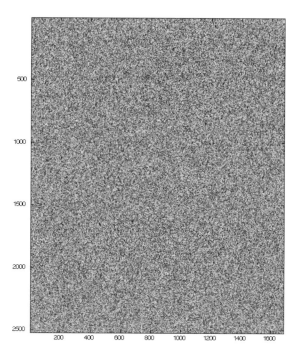

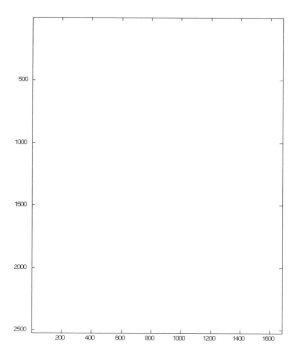

The encrypted message is y and the message is denoted by s, if the attacker chooses a black image for message, the signal s will be zero, and its encrypted message is $\overline{y} = Cx$, if a new message s_2 is transmitted it is easy to recover it simply by $s_2 = \overline{y} - Cx = Cx + s_2 - Cx$, note that finding the key is unnecessary.

A chosen plaintext attack can successfully retrieve the values that were used for encrypting the plain image, this is caused mainly because the encryption values will only change if the key is changed, so all messages will be encrypted with the same set of values producing the mentioned vulnerability. In Fig. 8.10 the result of the mentioned plaintext attack is shown.

Using as chosen message a black image allowed to recover the output of the system $\overline{y}(t) = Cx(t)$ without the influence of the message which is then used to decrypt future messages. This simple form of chaotic masking is not very effective when the data carrier signal is considerably different than the signals produced by the oscillators states, or has different periods and amplitudes that make the extraction of the signal easy.

Another form of encryption that relies on synchronization of fractional chaotic systems can be found on [32], the proposal found on this reference uses the synchronization of a Lorenz like system, then it obtains a sequence of pseudorandom numbers from the states of the system using an equation with fixed values, the image is encrypted by a bitwise XOR operation with the previously obtained sequence.

The mentioned encryption scheme causes the security of the data to depend solely on the inability of the attacker to synchronize to the master system, as the key is presumably used only to create the initial conditions and parameters of the Lorenz system. This is a considerable flaw as there are many ways to synchronize to the master system without the need of knowing its exact parameters nor initial conditions, for example the state observer that is introduced in this chapter can easily reconstruct its states without the exact parameters making a synchronization attack similar to [36] viable, the master system used for encryption is given by:

$$\dot{x} = f_1(x)$$
$$y = Cx$$

And a slight modification of the proposed observer can synchronize to it:

$$\dot{\hat{x}} = f_2(\hat{x}) + K \tanh(e)$$
$$\hat{y} = C\hat{x}$$

The functions f_1 and f_2 share the same structure but posses different values, because f_1 will have the key and to the attacker it is unknown, hence f_2 has not the exact same parameters, but as it was proven, the observer can reconstruct the states of the master even in the presence of disturbances (the message or an unknown key), once both systems have been synchronized all that remains is to create the encryption values, and as they depend

Fig. 8.10 Chosen plain image and resulting image of plaintext attack

solely on an equation that does not depend on the key or message, it is easy to compute them using the same equation, this finally allows the attacker to access the encrypted data without the need of the key. There are also other methods that could allow to compute the value of the key such as parametric estimation, but the proposed one is more practical. The implementation of the synchronization based cryptanalysis can be seen in Fig. 8.11.

The resulting image from the attack is very similar to the one used as message, it has a small difference in color in certain regions, this is caused by not using the exact same parameters, not knowing the key and the small error that the observer had. The encryption algorithm introduced in this chapter renders both mentioned forms of cryptanalysis useless, the message itself is not present in the output making the simple chosen plaintext attack ineffective, most importantly, by making the initial conditions depend on the message and the key, every signal used to encrypt the data will depend on the message itself, so each message will have a unique encryption causing that this type of attacks be useless. The synchronization attack will also be unsuccessful, because for the use of the bitwise XOR operation that depends on the key, so even if the attacker can achieve synchronization to the master system, the estimated data carrier signal will be useless without the key to the attacker. To test this a chosen plaintext attack will be implemented, a black image of size 1683×2522 pixels will be the chosen plaintext, the goal of the attack will be to recover Fig. 8.6, the chosen plaintext and the result of the attacks can be seen in Fig. 8.12.

The chosen plaintext attack failed to recover the message from Fig. 8.6, making evident that the encryption is unaffected by known and chosen plaintext attacks. A synchronization attack is performed next, the same technique that was used to recover a similar message from [32] will be applied, to make the attack possible some elements of the key are supposedly known, the parts of the key regarding the derivative order, and the timing of the data carrier signal are required so the interpretation of the data is easy to visualize, the attack produces the results shown in Fig. 8.13.

The attack is also ineffective, even if the observer can synchronize to the states, the mixing with pseudorandom numbers depending on the unknown key keeps the information well hidden, allowing the information to remain safe. The correlation of two adjacent pixels for the plain image in Fig. 8.6 and its encrypted message in Fig. 8.7 can be seen in Fig. 8.14.

The graphic reveals that the encryption is effective, as the plain image correlation of pixel is linear while the correlation of pixels in the encrypted image is not, finally the entropy of the encrypted image is 7.9951 which is very close to the ideal number 8, while not ideal, this indicates that there is not much predictability in the encrypted image, considering all this the encryption algorithm offers a good level of protection from the cryptographic techniques applied in this section and, according to the histograms and the correlation graphic, the proposed algorithm will also provide protection against linear, differential and other types of statistical cryptanalysis forms.

Fig. 8.11 Message and resulting image of synchronization attack

Fig. 8.12 Chosen plain image
and resulting image of
plaintext attack

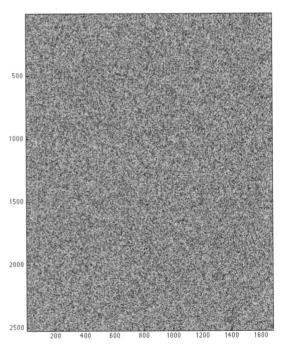

Fig. 8.13 Message and resulting image of synchronization attack

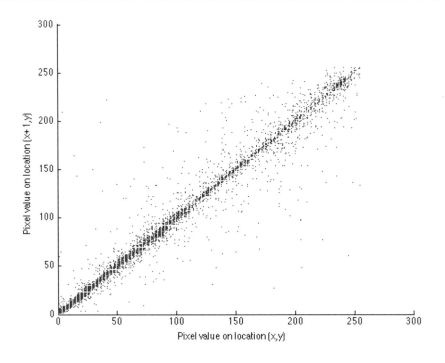

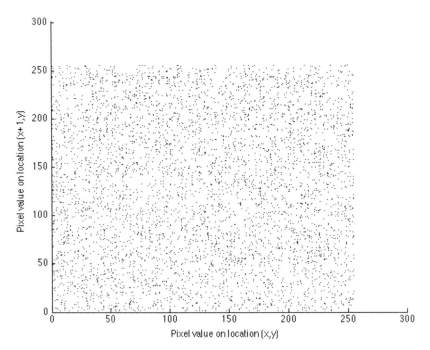

Fig. 8.14 Correlation of adjacent pixels of the plain image and the encrypted image

8.7   Concluding Remarks

Using the proposed fractional smoothed sliding modes observer made possible to recover data without any loss, Figs. 8.3, 8.4, and 8.5 show that the observer is capable of producing an accurate reconstruction of the states with negligible chattering, attaining this level of performance is crucial to maintain data recovery without loss, so the observer served its purpose well.

The algorithm was capable of recovering the messages without error, this can be seen on Figs. 8.6, 8.7, and 8.8, while having a reasonable execution time and linear time complexity when working with large color images. As previously stated, the algorithm focus is to provide security specially against cryptanalysis comprised of chosen plaintext attacks and known plaintext attacks, but it also posses resistance against linear and differential cryptanalysis as shown in the security analysis section.

To the best of the authors knowledge, the state of the art encryption algorithms that depend on fractional order chaotic systems synchronization are susceptible to cryptanalysis, in the security analysis section, it is shown that two of them can be easily broken by chosen plaintext attacks and synchronization based cryptanalysis, but, this work proved that it is possible to use fractional chaotic systems in an encryption algorithm while maintaining good performance in data reconstruction and security against the common cryptanalysis techniques that were mentioned.

References

1. Laskin, N. (2000). Fractional market dynamics. *Physica A: Statistical Mechanics and its Applications, 287*(3), 482–492.
2. Hilfer, R. (Ed.). (2000). *Applications of fractional calculus in physics.* World Scientific.
3. Zhou, X. J., Gao, Q., Abdullah, O., & Magin, R. L. (2010). Studies of anomalous diffusion in the human brain using fractional order calculus. *Magnetic Resonance in Medicine, 63*(3), 562–569.
4. Freeborn, T. J. (2013). A survey of fractional-order circuit models for biology and biomedicine. *IEEE Journal on Emerging and Selected Topics in Circuits and Systems, 3*(3), 416–424.
5. Martínez-Guerra, R., Pérez-Pinacho, C. A., Gómez-Cortés, G. C. (2015). Synchronization of chaotic Liouvillian systems: an application to Chua's oscillator. In *Synchronization of integral and fractional order chaotic systems.* Springer International Publishing.
6. Pecora, L. M., & Carroll, T. L. (1990). Synchronization in chaotic systems. *Physical Review Letters, 64*(8), 821.
7. Abanda, Y., & Tiedeu, A. (2016). Image encryption by chaos mixing. *IET Image Processing, 10*(10), 742–750.
8. Abd-El-Hafiz, S. K., Radwan, A. G., Haleem, S. H. A., & Barakat, M. L. (2014). A fractal-based image encryption system. *IET Image Processing, 8*(12), 742–752.
9. Kocarev, L., Halle, K. S., Eckert, K., Chua, L. O., & Parlitz, U. (1992). Experimental demonstration of secure communications via chaotic synchronization. *International Journal of Bifurcation and Chaos, 2*(03), 709–713.
10. Liao, T. L., & Tsai, S. H. (2000). Adaptive synchronization of chaotic systems and its application to secure communications. *Chaos, Solitons and Fractals, 11*(9), 1387–1396.

11. Liao, T. L., & Huang, N. S. (1999). An observer-based approach for chaotic synchronization with applications to secure communications. *IEEE Transactions on Circuits and Systems I: Fundamental Theory and Applications, 46*(9), 1144–1150.

12. Cuomo, K. M., Oppenheim, A. V., & Strogatz, S. H. (1993). Synchronization of Lorenz-based chaotic circuits with applications to communications. *IEEE Transactions on Circuits and Systems II: Analog and Digital Signal Processing, 40*(10), 626–633.

13. Feki, M. (2003). An adaptive chaos synchronization scheme applied to secure communication. *Chaos, Solitons and Fractals, 18*(1), 141–148.

14. Huang, X., Sun, T., Li, Y., & Liang, J. (2014). A color image encryption algorithm based on a fractional-order hyperchaotic system. *Entropy, 17*(1), 28–38.

15. Hsiao, H. I., & Lee, J. (2015). Color image encryption using chaotic nonlinear adaptive filter. *Signal Processing, 117*, 281–309.

16. Janakiraman, S., Thenmozhi, K., Rayappan, J. B. B., & Amirtharajan, R. (2018). Lightweight chaotic image encryption algorithm for real-time embedded system: Implementation and analysis on 32-bit microcontroller. *Microprocessors and Microsystems, 56*, 1–12.

17. Xu, L., Gou, X., Li, Z., & Li, J. (2017). A novel chaotic image encryption algorithm using block scrambling and dynamic index based diffusion. *Optics and Lasers in Engineering, 91*, 41–52.

18. Sun, S. (2017). Chaotic image encryption scheme using two-by-two deoxyribonucleic acid complementary rules. *Optical Engineering, 56*(11), 116117.

19. Teng, L., Wang, X., & Meng, J. (2017). A chaotic color image encryption using integrated bit-level permutation. *Multimedia Tools and Applications, 77*(6), 6883–6896.

20. Liu, H., Kadir, A., & Sun, X. (2017). Chaos-based fast colour image encryption scheme with true random number keys from environmental noise. *IET Image Processing, 11*(5), 324–332.

21. Liu, L., Miao, S., Hu, H., & Cheng, M. (2016). N-phase logistic chaotic sequence and its application for image encryption. *IET Signal Processing, 10*(9), 1096–1104.

22. TDridi, M., Hajjaji, M. A., Bouallegue, B., & Mtibaa, A. (2016). Cryptography of medical images based on a combination between chaotic and neural network. *IET Image Processing, 10*(11), 830–839.

23. Abd-El-Hafiz, S. K., Radwan, A. G., Haleem, S. H. A., & Barakat, M. L. (2014). A fractal-based image encryption system. *IET Image Processing, 8*(12), 742–752.

24. Wang, X. Y., & Gu, S. X. (2014). New chaotic encryption algorithm based on chaotic sequence and plain text. *IET Information Security, 8*(3), 213–216.

25. N'Doye, I., Darouach, M., & Voos, H. (2013, July). Observer-based approach for fractional-order chaotic synchronization and communication. In *European Control Conference (ECC)* (pp. 4281–4286).

26. Luo, C., & Wang, X. (2013). Chaos generated from the fractional-order complex Chen system and its application to digital secure communication. *International Journal of Modern Physics C, 24*(04), 1350025.

27. Wu, X., Wang, H., & Lu, H. (2012). Modified generalized projective synchronization of a new fractional-order hyperchaotic system and its application to secure communication. *Nonlinear Analysis: Real World Applications, 13*(3), 1441–1450.

28. Deng, Y. S., Qin, K. Y., & Shao, S. Q. (2009, July). Synchronization in coupled fractional order Chen-system and its application in secure communication. In *Communications, Circuits and Systems. ICCCAS 2009* (pp. 839–841).

29. Kiani-B, A., Fallahi, K., Pariz, N., & Leung, H. (2009). A chaotic secure communication scheme using fractional chaotic systems based on an extended fractional Kalman filter. *Communications in Nonlinear Science and Numerical Simulation, 14*(3), 863–879.

30. Sheu, L. J. (2011). A speech encryption using fractional chaotic systems. *Nonlinear Dynamics, 65*(1), 103–108.

31. Zhen, W., Xia, H., Ning, L., & Xiao-Na, S. (2012). Image encryption based on a delayed fractional-order chaotic logistic system. *Chinese Physics B*, *21*(5), 050506.

32. Xu, Y., Wang, H., Li, Y., & Pei, B. (2014). Image encryption based on synchronization of fractional chaotic systems. *Communications in Nonlinear Science and Numerical Simulation*, *19*(10), 3735–3744.

33. Goldreich, O. (2008). *Computational complexity: A conceptual perspective*. Cambridge University Press. 1st edition.

34. Martin, J. C. (1991). *Introduction to languages and the theory of computation* (Vol. 4). New York, NY: McGraw-Hill.

35. Alvarez, G., & Li, S. (2006). Some basic cryptographic requirements for chaos-based cryptosystems. *International Journal of Bifurcation and Chaos*, *16*(08), 2129–2151.

36. Yang, T., Yang, L. B., & Yang, C. M. (1998). Breaking chaotic switching using generalized synchronization: Examples. *IEEE Transactions on Circuits and Systems I: Fundamental Theory and Applications*, *45*(10), 1062–1067.

Secure Communications by Using Atangana-Baleanu Fractional Derivative

9

Abstract

In this chapter the authors propose to use fractional-order chaotic systems for data encryption, the encryption is a hybrid cipher, that takes elements of stream ciphers and block ciphers, to allow to handle large messages without compromising the security of the message or severely increasing the need of computational power to process the encryption algorithm. The cipher relies on the synchronization of chaotic systems by using state observers, the observer is capable of accurately recovering states and uncertainties within the fractional-order chaotic system. To test the algorithm and observer, the messages in this chapter are color images.

9.1 Introduction

Fractional calculus has seen developments in various fields like finance [1], physics [2], Medicine [3] and chaotic systems synchronization, in this last field, one of its most notorious applications are secure communications [4, 5], where a fractional-order chaotic system is used as a pseudorandom number generator, which in turn, serves as input for an encryption algorithm. The usual method employed for encryption is to mask a signal containing a message with a fractional-order chaotic oscillator state or output, the oscillator is named as the master system in synchronization, then the slave system task is to reconstruct the signal that contains the message and in consequence recover the encrypted message that the signal contains.

Since the introduction of synchronization by Pecora various methods have been used for the reconstruction of message signals, the most recurrently present in the literature are generalized synchronization, reconstruction of states by Liouvillian systems properties

© The Author(s), under exclusive license to Springer Nature Switzerland AG 2023
R. Martínez-Guerra et al., *Encryption and Decryption Algorithms for Plain Text and Images using Fractional Calculus*, Synthesis Lectures on Engineering, Science, and Technology, https://doi.org/10.1007/978-3-031-20698-6_9

and state observers [6], although most of them offer similar results in the precision of the reconstructed signal, the way in which the message is converted into a signal and how the signal is encrypted, impacts heavily on the performance of the decryption process, as in various cases the treatment given to the signal does not allow for any error in its reconstruction if the message is required to be correctly recovered, then it is of upmost importance to correctly choose how the master and slave systems are to be synchronized, and to craft the message signal so it properly fits into the synchronization and allows the correct recovery of messages.

In the literature there are several works on secure communications using fractional-order chaotic systems synchronization, in particular the most popular encryption are stream ciphers that relay on chaotic masking, but in the majority of cases it is not properly done and many crucial specifications on the message signal are missing. The preferred synchronization method are state observers, and the prevalent ones are variants of the Luenberger observer, this choice of observer limits where the message signal can be embedded in the master system, and in various ways, contributes to an inaccurate reconstruction of both the signal and the message.

In this chapter, the authors propose the implementation of a hybrid encryption scheme, that contains elements of chaotic masking, chaos shift keying and block ciphers that can guarantee an error free message reconstruction even if the reconstruction of the signal is not totally accurate and with high tolerance to noise within the message signal. To achieve the desired result an uncertainty state observer for fractional-order chaotic systems is proposed as the slave system, the set of pseudorandom numbers is generated through a fractional-order chaotic oscillator with Mittag-Leffler kernel [7, 8] and a variation of the Blum Blum Shub generator.

9.2   Preliminaries

The proposed work is based on the following definitions:

Definition 9.1 The Riemann-Liouville fractional derivative of a function $f(t)$ is:

$$^{RL}D_t^\alpha f(t) = \frac{1}{\Gamma(n-\alpha)} \int_0^t \frac{f^n(\tau)}{(t-\tau)^{n-\alpha+1}} d\tau$$

With $n-1 < \alpha < n$ and $\Gamma(\bullet)$ is the gamma function given by $\Gamma(\alpha) = \int_0^\infty t^{\alpha-1} e^{-t} dt$ that converges to the right half of the complex plane.

Definition 9.2 The Riemann-Liouville fractional integral of a function $f(t)$ is:

$$_0I_t^\alpha f(t) = \frac{1}{\Gamma(\alpha)} \int_0^t f(\tau)(t-\tau)^{\alpha-1} d\tau$$

Definition 9.3 The Caputo fractional derivative of order α of a function $f(t)$ is described by:

$$_{t_o}^{C}D_t^{\alpha} f(t) = \frac{1}{\Gamma(n-\alpha)} \int_{t_o}^{t} f^{(n)}(\tau)(t-\tau)^{n-\alpha-1} d\tau$$

And $f^{(n)}(\tau)$ is the n-th derivative, n is positive integer number.

Recent developments in definitions of fractional derivatives correct some of the issues of the convolution law based fractional-order derivatives, such as Riemman-Liouville and Caputo, these problems involve mostly singularities, one of this new derivative definitions is given next:

Definition 9.4 For $f \in H^1(a,b)$, $a < b$ and $\alpha \in [0,1]$ The Atangana-Baleanu fractional derivative of order α of a function $f(t)$ is described by:

$$^{ABC}D_t^{\alpha} f(t) = \frac{W(\alpha)}{1-n} \int_a^t f^{'}(\tau) E_{\alpha}\left[-\alpha \frac{(t-y)^{\alpha}}{1-\alpha}\right] d\tau$$

Where $W(\bullet)$ with $W(0) = W(1)$ is a normalization function and E_{α} is the Mittag-Leffler function defined by:

$$E_{\alpha}(z) = \sum_{n=0}^{\infty} \frac{z^n}{\Gamma(n\alpha+1)}$$

Lemma 9.1 *The Atangana-Baleanu Fractional differential equation $_0^{ABC}D_t^{\alpha} f(t) = u(t)$ has a unique solution if the equation fulfills [7]:*

$$f(t) = \frac{1-\alpha}{W(\alpha)} u(t) + \frac{\alpha}{W(\alpha)\Gamma(\alpha)} \int_0^t u(\tau)(t-\tau)^{\alpha-1} d\tau$$

Lemma 9.2 *If a real positive definite function $V[x(t)]$ exists, and it results to a negative semi-definitie fractional function defined by $D^{\alpha}[x(t)]$ with $\alpha \in (0,1]$, then the fractional system is asymptotically stable [7,9] this is:*

$$a_1 \|x\| \leq V(x) \leq a_2 \|x\|$$

$$D^{\alpha} V(x) \leq -a_3 \|x\|$$

With $a_1, a_2, a_3 > 0$.

9.3   Encryption Algorithm

The encryption algorithm is similar to chaos shift keying, in the sense that it will employ a binary data carrier signal $g(s, y)$ to change the value of one of the parameters of the system, but with the difference that it does not change the attractors shape, it rather just behaves like an unknown parameter, the dynamic of the transmitter is given next:

$$^{ABC}_{0}D_t^\alpha x_1 = Ax + f(x) + C^T s_c$$

$$y = Cx$$

Where $0 < \alpha < 1$, the function $f(x)$ is bounded by the Lipschitz condition:

$$\|f(w) - f(z)\| \leq \Gamma \|w - z\|, \ \Gamma, w, z \in \mathbb{R}$$

The set of instructions for creating the data carrier signal and parameters of the system is presented next, for it, it is assumed that the data is a color image of size ij composed of three matrices that contain the primary color intensity of the image denoted by $R \in \mathbb{N}^{i \times j}$, $0 \leq R_{ij} \leq 255$, $G \in \mathbb{N}^{i \times j}$, $0 \leq G_{i \times j} \leq 255$ and $B \in \mathbb{N}^{i \times j}$, $0 \leq B_{ij} \leq 255$.

1. The key is a set of decimal values separated into smaller sections denoted Key_n, $n = 1, 2, 3, \ldots$, the key must contain enough elements for the subsequent steps

$$Key = 12345 - 12345 - 12345 - 12345 - 12345 - \ldots$$

$$Key = key_1 - key_2 - key_3 - key_4 - key_5 - \ldots$$

2. The initial conditions of the oscillator are given by the key and the message, the next equations describe the process:

$$x_1(0) = \frac{Key_1}{99999} \frac{\sum_{a=1}^{ij} R_a}{255ij}, \ x_2(0) = \frac{Key_2}{99999} \frac{\sum_{a=1}^{ij} G_a}{255ij}$$

The remaining parameters can be computed by only using the key, for example the value of $\alpha \in [0, 1]$ is obtained by:

$$\alpha = \frac{key_3}{99999}$$

3. The data carrier signal s is made from the message, since the image is formed by 8 bit integers they have to be converted into its binary representation, then arranged into a single vector of size $24ij$ and then subtract 0.5 so that a positive period is a 1 and a

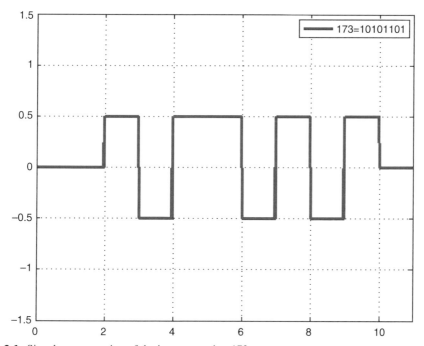

Fig. 9.1 Signal representation of the integer number 173

negative is 0, the period of the signal is given by:

$$P = \frac{Key_4}{99999}$$

For example, Fig. 9.1 shows how to convert the integer 173 into its equivalent binary signal with a 1 s period.

4. Factorize two key elements in prime numbers, the two largest primes p and q are the inputs for the pseudo random number generator given by the Blum Blum Shub algorithm:

$$r_{n+1} = r_n^2 \mod (pq)$$

The number of iterations for the Blum Blum Shub algorithm is given by another element of the key.

5. The encrypted signal is described by the equation:

$$s_c = s \oplus r$$

Where $\oplus$ is the bitwise XOR operation.

By using this algorithm the problem of the encrypted message not depending on the transmitted message has been addressed, even more, using the binary signal will allow a certain room for error in the reconstruction while not compromising the integrity of the data, as it is only needed to be accurate on the sign of the reconstructed data carrier signal, unlike chaos masking that must have perfect reconstructions to guarantee no data loss.

9.4   Receiver and Decryption

The receiver will be a sliding modes state observer, this kind of observer can easily deal with unknown parameters such as the binary data carrier signal, but also the binary signal can maintain zero data loss in the presence of chattering, making it appropriate for this task, the observer dynamic is described:

$$
{}^{ABC}_{0}D^{\alpha}_{t}\hat{x} = A\hat{x} + f\left(\hat{x}\right) + k_a\left(y - \hat{y}\right) + k_b\tanh\left(y - \hat{y}\right)
$$
$$
{}^{ABC}_{0}D^{\alpha}_{t}\gamma = k_c\tanh\left(y - \hat{y}\right)
$$
$$
\hat{y} = C\hat{x}
$$
$$
\hat{s}_c = {}^{ABC}_{0}D^{\alpha}_{t}y - f\left(\hat{x}\right)
$$

The synchronization error is:

$$
e = x - \hat{x}
$$

The error's fractional-order derivative of order α is:

$$
{}^{ABC}_{0}D^{\alpha}_{t}e = Ae + \phi\left(e\right) + C^T s - K_a Ce - K_b\tanh(Ce)
$$

Where $\phi\left(e\right) = f\left(x\right) - f\left(\hat{x}\right)$ and $Cs < \Lambda$ is bounded by the positive amplitude of the binary signal.

Assumption 9.1 *The next LMI has solutions $P = P^T > 0$ and $Q = Q^T > 0$:*

$$
\left(A - K_a C\right)^T P + P\left(A - K_a C\right) \leq -Q
$$

Assumption 9.2 *The message signal and the nonlinear part of the oscillator are bounded:*

$$
\|s_c\| \leq S_{max}
$$
$$
\|f\left(x\right) - f\left(y\right)\| \leq \Phi\|x - y\|
$$
$$
2xP\left(C^T S_{max} + \Phi_v\right) \leq \Lambda\|x\|
$$

For a positive real number S_{max} and Λ, and a vector Φ_v that contains the bounds Φ for the non linear part of each state.

A function that fulfills Lemma 9.2 is:

$$V = e^T P e$$

$${}^{ABC}_0 D^\alpha_t V = (D^\alpha e)^T Pe + eP^T (D^\alpha e)$$

$$+2 \sum_{k=0}^{\infty} \frac{\Gamma(1+\alpha)(D^k e)^T (D^{\alpha-k} e)}{\Gamma(1+k)\Gamma(1-k+\alpha)}$$

Knowing that $D^n x$ with $n = 1, 2, 3 \ldots$ exists and is continuous and bounded, there is an $N \in \mathbb{R}$, $N \geq 0$ such that $\|D^n x\| \leq N$, for a real non integer α there is an integer U such that $U - 1 < \alpha < U$, and the derivative $D^{\alpha-n} x$ is divided into two parts: $D^{\alpha-n} x, n = 1, 2, \ldots, U - 1$ and $D^{\alpha-n} e, n = U, U+1, U+2, \ldots$ for the fist part $D^{\alpha-n} x$ for $n = 1, 2, \ldots, U - 1$ there is a real number $H > 0$ that makes $\left\| D^{\alpha-n} x \right\| = \left\| I^{\alpha-n} x \right\| \leq H \|x\|$, the fractional-order nonlinear system can be expressed as $D^\alpha x = f(x)$. The dynamic of the system is bounded by $\|f(x)\| < J \|x\|$ the inequality $\left\| D^{\alpha-n} x \right\| \leq HJ \|x\|$ is true, for the next part $D^{\alpha-n} e, n = U, U+1, U+2, \ldots$ by a similar process, it is obtained that $\left\| D^{\alpha-n} x \right\| \leq H_{max} \|x\|$ for a real number $H_{max} > 0$, so $\left\| D^{\alpha-n} x \right\| \leq \bar{H} \|x\|$ with $\bar{H} = max\{HJ, H_{max}\}$, with the next inequality $0 < J_{min} < |\Gamma(1 - \alpha + n)|$, for a positive real number $J_{min} > 0$ and since $\frac{\Gamma(n)}{\Gamma(n+1)} = \frac{1}{n}$ for $n = 1, 2, 3, \ldots$, the expression $\sum_{n=1}^{\infty} \frac{1}{\Gamma(1+n)}$ converges, so there is a $K > 0$ that $0 < \sum_{n=1}^{\infty} \frac{1}{\Gamma(1+n)} < K$, making the Lyapunov fractional-order derivative:

$${}^{ABC}_0 D^\alpha_t V \leq \left({}^{ABC}_0 D^\alpha_t e\right)^T Pe + e^T P \left({}^{ABC}_0 D^\alpha_t e\right) + 2B \|e\|$$

Making $B = \frac{\Gamma(1+\alpha)NK\bar{H}}{J_{min}}$ and $\tanh(x) \leq sign(x) + \gamma$ for an appropriate positive real number γ:

$${}^{ABC}_0 D^\alpha_t V \leq e^T \left[(A - K_a C)^T P + P(A - K_a C) \right] e$$

$$+2e^T P \left[C^T s_c + \phi(e) - K_b sign(Ce) - K_b \gamma \right] + 2B \|e\|$$

By Assumption 9.2:

$${}^{ABC}_0 D^\alpha_t V \leq e^T \left[(A - K_a C)^T P + P(A - K_a C) \right] e$$

$$+2e^T P \left(C^T S_{max} + \Phi_v \right) - 2e^T P K_b sign(Ce) - 2e^T P k_b \gamma + 2B \|e\|$$

$$\leq -e^T Q e + 2\Lambda \|e\| - 2e^T P K_b sign(Ce) - 2e^T P k_b \gamma + 2B \|e\|$$

The inequality $2e^T P k_b \gamma < \varepsilon \|e\|$ leads to:

$$\begin{smallmatrix}ABC\\0\end{smallmatrix} D_t^\alpha V \le -e^T Q e + 2 (\Lambda - \varepsilon + B) \|e\| - \frac{2}{\sqrt{\lambda_{max}(C^T C)}} e^T P K_b C \frac{e}{\|e\|}$$

So:

$$\begin{smallmatrix}ABC\\0\end{smallmatrix} D_t^\alpha V \le 2[(\Lambda - \varepsilon + B) - \lambda_{min}(P K_b C)] \|e\| - 2 e^T P \varepsilon$$

Making K_b such that $\lambda_{min}(P K_b C) > \Lambda + B - \varepsilon$ yields:

$$\begin{smallmatrix}ABC\\0\end{smallmatrix} D_t^\alpha V \le 0$$

With this last result it is possible to conclude that the synchronization error will be Mittag-Leffler stable, hence it makes possible to reconstruct the states and recover the message by:

$$\hat{s}_c = \begin{smallmatrix}ABC\\0\end{smallmatrix} D_t^\alpha y - f(\hat{x})$$

Having recovered the estimated data carrier signal, the key is used to reconstruct the values of the binary signal and the pseudorandom numbers used to encrypt it. The resulting message is the arranged into the recovered message.

9.5   Numerical Results

Stream ciphers can handle large data much better than block ciphers, to show the capabilities of the proposed encryption algorithm a simulation will be implemented using the next fractional-order Rossler oscillator as master system:

$$\begin{smallmatrix}ABC\\0\end{smallmatrix} D_t^\alpha x_1 = -x_2 - x_3$$
$$\begin{smallmatrix}ABC\\0\end{smallmatrix} D_t^\alpha x_2 = x_1 + a x_2 + s$$
$$\begin{smallmatrix}ABC\\0\end{smallmatrix} D_t^\alpha x_2 = b + x_3 (x_1 - r)$$
$$y = x_2$$

The Rossler fractional-order chaotic oscillator has the kernels defined by:

$$\mathcal{K}_1 (x_1, t) = -x_2 - x_3$$
$$\mathcal{K}_2 (x_2, t) = x_1 + a x_2 + s$$
$$\mathcal{K}_3 (x_3, t) = b + x_3 (x_1 - r)$$

Using Lemma 9.1, the kernels $\mathcal{K}_i$, $i \in 1, 2, 3$ lead to the following equation of the states:

$$x_1(t) = x_1(0) + \frac{1 - \alpha}{W(\alpha)} \mathcal{K}_1(x_1, t) + \frac{\alpha}{W(\alpha)\Gamma(\alpha)} \int_0^t (t - y)^{\alpha - 1} \mathcal{K}_1(y, x_1)\, dy$$

$$x_2(t) = x_2(0) + \frac{1 - \alpha}{W(\alpha)} \mathcal{K}_2(x_2, t) + \frac{\alpha}{W(\alpha)\Gamma(\alpha)} \int_0^t (t - y)^{\alpha - 1} \mathcal{K}_2(y, x_2)\, dy$$

$$x_3(t) = x_3(0) + \frac{1 - \alpha}{W(\alpha)} \mathcal{K}_3(x_3, t) + \frac{\alpha}{W(\alpha)\Gamma(\alpha)} \int_0^t (t - y)^{\alpha - 1} \mathcal{K}_3(y, x_3)\, dy$$

A method for the numerical approximation for the states of a fractional-order system, using the Atangana-Baleanu derivative is given in [10], the equation for the approximation of a state depending on its kernel is:

$$u_{n+1} = u_0 + \frac{1 - \alpha}{W(\alpha)} f[u(t_n), t_n]$$

$$+ \frac{\alpha}{W(\alpha)} \sum_{s=0}^{n} \left\{ \frac{h^\alpha f(u_s, t_s)}{\Gamma(\alpha + 2)} \left[(n + 1 - s)^\alpha (n - s + 2 + \alpha) \right. \right.$$

$$- (n - s)^\alpha (n - s + 2 + 2\alpha)]$$

$$\left. - \frac{h^\alpha f(u_{s-1}, t_{s-1})}{\Gamma(\alpha + 2)} \left[(n + 1 - s)^\alpha - (n - s)^\alpha (n - s + 1 + \alpha) \right] \right\}$$

$$+ E_n^\alpha$$

Where $E_n^\alpha = \frac{\alpha}{W(\alpha)\Gamma(\alpha)} \sum_{s=0}^{n} \int_{t_s}^{t_{s-1}} \frac{(y - t_s)(y - t_{s-1})}{2} \frac{\partial^2}{\partial y^2} [f(u(y), y)]_{y=\lambda_y} (t_{n+1} - y)^{\alpha - 1}\, dy$, making the states for the numerical simulation be computed by the next equations:

$$x_{1,n+1} = x_{1,0} + \frac{1 - \alpha}{W(\alpha)} \mathcal{K}_1[x_1(t_n), t_n]$$

$$+ \frac{\alpha}{W(\alpha)} \sum_{s=0}^{n} \left\{ \frac{h^\alpha \mathcal{K}_1(x_{1,s}, t_s)}{\Gamma(\alpha + 2)} \left[(n + 1 - s)^\alpha (n - s + 2 + \alpha) \right. \right.$$

$$- (n - s)^\alpha (n - s + 2 + 2\alpha)]$$

$$\left. - \frac{h^\alpha \mathcal{K}_1(x_{1,s-1}, t_{s-1})}{\Gamma(\alpha + 2)} \left[(n + 1 - s)^\alpha - (n - s)^\alpha (n - s + 1 + \alpha) \right] \right\}$$

$$+ E_n^\alpha$$

$$x_{2,n+1} = x_{2,0} + \frac{1-\alpha}{W(\alpha)} \mathcal{K}_2 [x_2(t_n), t_n]$$

$$+ \frac{\alpha}{W(\alpha)} \sum_{s=0}^{n} \left\{ \frac{h^\alpha \mathcal{K}_2(x_{2,s}, t_s)}{\Gamma(\alpha+2)} \left[(n+1-s)^\alpha (n-s+2+\alpha) \right. \right.$$

$$- (n-s)^\alpha (n-s+2+2\alpha) \Big]$$

$$- \frac{h^\alpha \mathcal{K}_2(x_{2,s-1}, t_{s-1})}{\Gamma(\alpha+2)} \left[(n+1-s)^\alpha - (n-s)^\alpha (n-s+1+\alpha) \right] \Bigg\}$$

$$+ E_n^\alpha$$

$$x_{3,n+1} = x_{3,0} + \frac{1-\alpha}{W(\alpha)} \mathcal{K}_3 [x_3(t_n), t_n]$$

$$+ \frac{\alpha}{W(\alpha)} \sum_{s=0}^{n} \left\{ \frac{h^\alpha \mathcal{K}_3(x_{3,s}, t_s)}{\Gamma(\alpha+2)} \left[(n+1-s)^\alpha (n-s+2+\alpha) \right. \right.$$

$$- (n-s)^\alpha (n-s+2+2\alpha) \Big]$$

$$- \frac{h^\alpha \mathcal{K}_3(x_{3,s-1}, t_{s-1})}{\Gamma(\alpha+2)} \left[(n+1-s)^\alpha - (n-s)^\alpha (n-s+1+\alpha) \right] \Bigg\}$$

$$+ E_n^\alpha$$

The receiver will be the presented sliding modes state observer:

$$_0^{ABC} D_t^\alpha \hat{x}_1 = -\hat{x}_2 - \hat{x}_3 + k_1 e + k_2 \tanh(e) + \gamma$$

$$_0^{ABC} D_t^\alpha \hat{x}_2 = \hat{x}_1 + a\hat{x}_2 + k_3 e + k_4 \tanh(e) + \gamma$$

$$_0^{ABC} D_t^\alpha \hat{x}_2 = b + \hat{x}_3(\hat{x}_1 - r) + k_5 e + k_6 \tanh(e) + \gamma$$

$$_0^{ABC} D_t^\alpha \gamma = \begin{bmatrix} k_7 \\ k_8 \\ k_9 \end{bmatrix} \tanh(e)$$

$$y = \hat{x}_2$$

$$\hat{s} = _0^{ABC} D_t^\alpha y - \hat{x}_1 - a\hat{x}_2$$

The next parameters will lead to chaotic behavior $a = 0.2$, $b = 0.2$, $r = 7$ and gains are $k_a = \begin{bmatrix} -3.485 \ 5.978 \ -2.866 \end{bmatrix}^T$, $k_b = \begin{bmatrix} 6.204 \ 8.715 \ -3.624 \end{bmatrix}^T$ and $k_a = \begin{bmatrix} -7.395 \ 6.925 \ -3.147 \end{bmatrix}^T$, since a stream cipher can handle large data a 12 megapixel

Fig. 9.2 Original message

(4032×3024) color image is an adequate example. In Fig. 9.2 the results of the encryption an decryption process are presented along the original message.

The encryption process is adequate, as it is possible to see (Fig. 9.3), there are no traces of the message image into the encrypted message. The performance of the state observer can be seen in Figs. 9.4, 9.5, 9.6, 9.7, and 9.8.

The observer can easily reconstruct the message and the states of the transmitter, while achieving zero data loss, because of the combination of a binary data carrier signal with a robust state observer, the proposed algorithm made use of the binary signal used in chaos shift keying, to avoid the data loss that any error in the reconstruction of the states would cause, using tanh instead of sign, attenuated the sliding modes chattering which would have led to reconstruction error if the encryption was done by a simple chaotic masking.

9.6 Security Analysis

The current examples of stream ciphers in the literature that use fractional-order chaotic systems, fail to provide security against cryptanalysis based on chosen plaintext attacks and known plaintext attacks, the known plaintext attack consists in the attacker knowing several messages and the corresponding encrypted message, which are used to retrieve information that will allow to decrypt new encrypted messages by identifying the key or

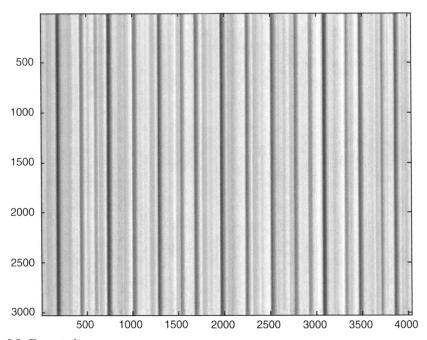

Fig. 9.3 Encrypted message

Fig. 9.4 Recovered message

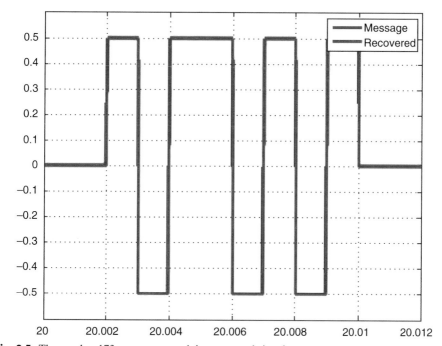

Fig. 9.5 The number 173 as message and the recovered signal

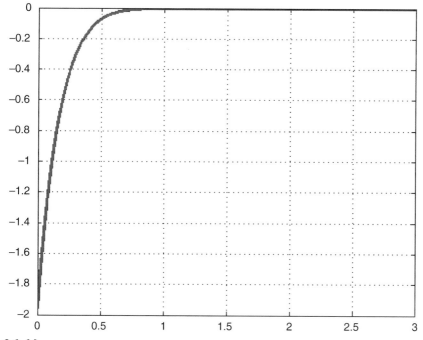

Fig. 9.6 Message recovery error

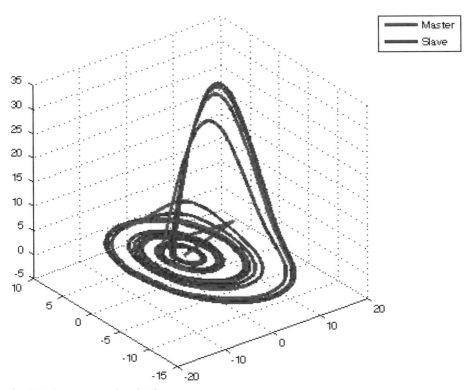

Fig. 9.7 Attractor synchronization

generating a equivalent key, a chosen plaintext attack is similar to the last one, with the difference that the attacker has access to the encryption device and can send specifically designed messages that will yield encrypted messages that contain valuable information to also recover the key or generate an equivalent key that will allow unauthorized decryption of further messages.

Stream ciphers used in the literature can be described by the next equation:

$$^{ABC}_0 D^\alpha_t x = f(x)$$

$$y = Cx + s$$

A practical and very easy way to break this encryption is to use a chosen plaintext attack, since the revised literature does not specify any way to make the encrypted message dependent on the message or simply assign the initial condition to fixed key values, generating an equivalent key can be done by designing a first message that causes $s = 0$, in a common masking a black image should do it, then the recovered message will be $y = Cx$, being Cx the equivalent key, as the states only change if the key changes, it will always be the same regardless of the message, then, the following messages can easily be retrieved with the equivalent key: $s_2 = y - Cx = Cx + s_2 - Cx$.

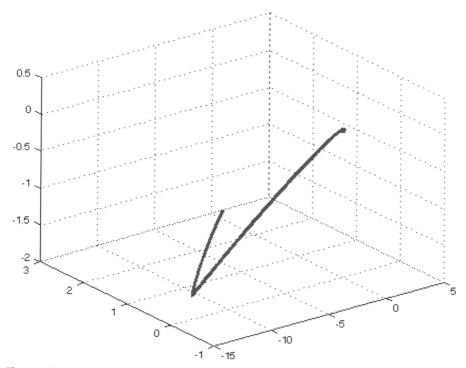

Fig. 9.8 Synchronization error

The proposed encryption algorithm avoids this problem by making the encrypted message dependent on the message, even more, the data carrier signal is not directly present on the encrypted message, so separating it from the rest of the message would require knowledge of several elements of the key that were used for the system's parameters, to test this a chosen plaintext attack will be implemented. A black 12 megapixels image is used as chosen message with the intention to recover the message used in the numerical results section, the result of the attack is shown in Fig. 9.9.

The attack clearly is unsuccessful, since the signals used to encrypt the black image are completely different than the ones used for the original message, the equivalent key is useless, showing that it is possible to make a safe stream cipher while using fractional-order chaotic systems.

9.7 Concluding Remarks

The encryption algorithm shows that it is possible to implement a secure stream cipher while using fractional-order chaotic systems. The sliding modes state observer along with the binary signal made possible to operate without data loss and removing the chattering

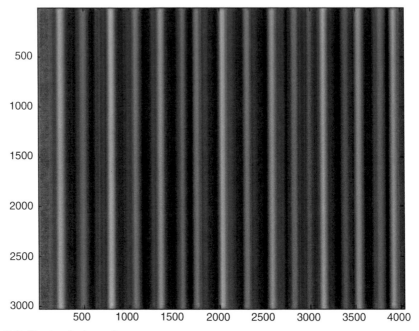

Fig. 9.9 Cryptanalysis result

that sliding modes posses, by correctly using the properties of fractional-order chaotic systems, the issue of not making the encrypted message depend on the message was solved, making the proposed stream cipher able to have no data loss and help the data remain safe.

The current works present on the literature about secure communications on fractional-order chaotic systems did not allow to safely transmit data while guaranteeing the integrity of the message, therefore, this result shows an alternative that can solve both problems while remaining fast and able to manage large quantities of data.

References

1. Laskin, N. (2000). Fractional market dynamics. *Physica A: Statistical Mechanics and its Applications, 287*(3–4), 482–492.
2. Hilfer, R. (2000). *Applications of fractional calculus in physics*. Singapore: World Scientific.
3. Zhou, X. J., Gao, Q., Abdullah, O., & Magin, R. L. (2010). Studies of anomalous diffusion in the human brain using fractional order calculus. *Magnetic Resonance in Medicine, 63*(3), 562–569.
4. Montesinos-Garcia, J. J., & Martinez-Guerra, R. (2019). A numerical estimation of the fractional-order Liouvillian systems and its application to secure communications. *International Journal of Systems Science, 50*(4), 791–806.
5. Montesinos-Garcia, J. J., & Martinez-Guerra, R. (2018). Colour image encryption via fractional chaotic state estimation. *IET Image Processing, 12*(10), 1913–1920.

6. Martinez-Guerra, R., Perez-Pinacho, C. A., & Gomez-Cortes, G. C. (2015). Synchronization of integral and fractional order chaotic systems. In *A differential algebraic and differential geometric approach*. Berlin: Springer.
7. Owolabi, K. M., Gomez-Aguilar, J. F., & Karaagac, B. (2019). Modelling, analysis and simulations of some chaotic systems using derivative with Mittag–Leffler kernel. *Chaos, Solitons & Fractals, 125*, 54-63.
8. Goufo, E. F. D., & Atangana, A. (2019). Modulating chaotic oscillations in autocatalytic reaction networks using atangana–baleanu operator. In *Fractional derivatives with Mittag-Leffler kernel* (pp. 135–158). Cham: Springer.
9. Li, Y., Chen, Y., & Podlubny, I. (2009). Mittag–Leffler stability of fractional order nonlinear dynamic systems. *Automatica, 45*(8), 1965–1969.
10. Toufik, M., & Atangana, A. (2017). New numerical approximation of fractional derivative with non-local and non-singular kernel: Application to chaotic models. *The European Physical Journal Plus, 132*(10), 444.

Index

A
ASCII, 48
Asymmetric key, 2
Asymptotically stable, 12, 161
Atangana-Baleanu, 223
Autonomous system, 13

B
Basins of attraction, 14
Binary number, 49, 168, 196
Bitwise XOR operation, 6, 56
Block cipher, 6, 83, 159, 222, 228
Block encryption, 5
Blum Blum Shub, 73

C
Chua oscillator, 38, 77
Cipher, 2, 60, 129, 184, 208, 234
Ciphertext, 47, 113, 118
Color image, 69, 118, 201
Colpitts oscillator, 81, 119
Cryptanalysis, 5, 130, 165, 192, 231
Cryptographic, 2, 57, 187, 212
Cryptography, 49
CSK, 1
CTAK, 72

D
Decryption, 2, 47, 127, 160, 192, 222
Duffing oscillator, 13, 75, 201

E
Embedded, 80, 194
Encrypted message, 3, 89, 123, 179, 205
Encryption, 1, 47, 117, 191, 224
Equilibrium point, 9, 161
Euler's number, 141
Exponential polynomial, 123

F
Fractal dimension, 41
Fractional-order, 151, 217
Fractional-order chaotic system, 221

G
Gamma function, 134, 161, 198, 222
Gauss error integral, 136
Grayscale image, 63

H
Histogram, 129, 171, 201

I
Instability theorem, 28

J
Julia set, 41, 118

Printed in the United States
by Baker & Taylor Publisher Services